T0211440

PLC Programming In Instruction List According To IEC 61131-3

Hans-Joachim Adam · Mathias Adam

PLC Programming In Instruction List According To IEC 61131-3

A Systematic And Action-Oriented Introduction In Structured Programming

 Springer

Hans-Joachim Adam
Bühl, Germany

Mathias Adam
Bühl, Germany

ISBN 978-3-662-65253-4   ISBN 978-3-662-65254-1   (eBook)
https://doi.org/10.1007/978-3-662-65254-1

Preface to the Fourth Edition (german)

This book is the result of teaching practice at vocational and general schools, in-house training and further education in a chemical company and the professional qualification of engineers to teach information technology at vocational schools in Baden-Württemberg (Germany).

The book is written as a textbook and exercise book. This means that you will be introduced to new areas through explanation or examples. To consolidate your knowledge and to check your learning and success, we recommend that you do the exercises. Sample solutions are available for most exercises: Each exercise has a unique name under which the solution can be found on the authors'[1] website.

An integral part of this book is the PLC simulation software "PLC-lite", which was developed by the authors especially for this course. You can download it free of charge from the authors' website. This gives you the opportunity to intensively test a control system that conforms to the standards, regardless of limitations with real systems. The scope of commands, data types, structures, etc. has been selected in such a way that the special features of programming in instruction lists according to IEC 61131-3 can be easily understood using the exercises from this book. In addition, "PLC-lite" contains a large number of processes as animated simulations. This allows you to test your programs in a practical and safe manner and to study the behavior of the controlled processes (even under extreme conditions or "errors"). Your PC is thus both a programming device and PLC automation device as well as a "technical system".

At the beginning we will deal in detail with basics and digital technology in the first four chapters. This part serves primarily as preparation for PLC programming, which is covered in the second part. You will learn important basics for PLC technology here. In addition, digital technology is already PLC programming! In principle, nothing else is done in the "function block diagram" language than connecting the logical symbols to form the program. This looks very similar to creating a digital circuit.

---

[1] www.adamis.de/plc/

In the second part of the book we cover PLC technology. You will now quickly see the advantage of PLC programming: Changes don't require cumbersome cable pulling, as was necessary with digital circuitry. You simply write the new instruction and that's it.

In this book, programming is learned with the most universal language, the *instruction list* (*IL*). Compared to the often more popular graphical languages, it has the advantage of an exact, easily comprehensible and clear structure. With the second edition, IEC 61131–3 is even more consistently oriented towards general programming languages.

With the book "SPS Programmierung mit IEC 61131" ("PLC Programming with IEC 61131") Karl Heinz John and Michael Tiegelkamp have created a reference that provides deep insights into the IEC 61131 standard. With this reference work, the fundamentals laid in this book can be systematically expanded.

Because this book is a textbook (and not a reference book), it is advisable to work through the book systematically. The subject matter and the exercises are arranged according to didactic aspects in a very specific order, building on each other. Please do not skip any of the exercises and only continue working when you have fully understood them, so that any gaps do not become noticeable later in a negative way. The step-by-step procedure guarantees the highest possible learning success.

To work through it, we recommend one of the following three methods:

The thorough method	You work through everything in sequence, chronologically from the first page to the last. If you take the time and have the patience to work through the more than 150 exercises, you will gain solid knowledge in both digital technology and PLC that goes far beyond simple basics. Especially if you have no prior knowledge of digital technology, this is the recommended approach. You will then learn digital technology first and then PLC technology
The parallel method	Alternatively, you can also learn digital technology and PLC technology at the same time by alternating between a chapter on digital technology and the corresponding PLC chapter. We recommend this procedure if you already know some digital technology, but are no longer completely sure. In this case, work through the chapters in the following order: 1, 2, 5, 3, 6, 7, 4, 8, 9, 10, 11, 12
The advanced method	If you are a digital technology "pro" and want to deal exclusively with PLC technology, you can simply start with the fifth chapter and then continue to the end of the book

Since the first edition of this book more than 13 years ago, the IEC 61131 standard has gained enormously in importance. The currently valid second edition of the standard brought numerous changes and additions, which have been taken into account in this fourth edition of the book. The flowchart illustrations in this book have been redrawn in accordance with DIN EN 60848.

Bühl, Germany   Hans-Joachim Adam
Spring 2012   Mathias Adam

Preface to the Fifth, Expanded Edition

We have added another chapter to the book in which the control of a mountain railway is programmed. Of particular interest here is the visualization of the mountain railway in the newly designed PLC-lite simulation application. This allows you to run the cable car "properly" and test your programs realistically and safely. By simulating operating errors (interrupted or short-circuited sensor) you can test the stability of your solutions. Furthermore, we have extended the subject index by a listing of the exercise tasks.

Bühl, Germany   Hans-Joachim Adam
Spring 2015   Mathias Adam

Contents

About the Authors

Hans-Joachim Adam (Graduate engineer) studied electrical engineering at the University of Karlsruhe. Since 1978, he has been teaching mathematics, electrical engineering, and information technology at the Technical High School in Bühl. At the State Seminar for Didactics and Teacher Training (vocational schools) in Karlsruhe, he is head of the information technology department. He is active in teacher training at the Oberschulamt Karlsruhe and the Kultusministerium Baden-Württemberg as well as in in-house training at various industrial companies.

Mathias Adam (Graduate engineer) studied electrical engineering and information technology at the University of Karlsruhe and works as a freelance consulting engineer. His work focuses on machine vision and embedded Linux.

Part I

Digital Technology

In the first part of the book we cover digital technology. This section serves as preparation for PLC programming, which is covered in Part II.

You will learn important basics for PLC technology. In addition, digital technology is already PLC programming! In the language "function block diagram" (which is not used in this book, however), basically nothing else is done than connecting the digital symbols to the circuit (and thus to the program).

Basics: Number Systems, Dual Numbers and Codes

1

Abstract

For most people today, the decimal number system commonly used in counting seems to be inherently given. It is seldom regarded as a pure invention of man. However, once the problems of counting and number systems are considered, the structure and necessity of number systems other than decimal become understandable. In Sects. 1.1, 1.2, 1.3, 1.4, 1.5, 1.6, 1.7, 1.8, 1.9 and 1.10 you will learn something about numbers and digits, what "counting" means and according to which laws numbers are constructed. The problem of coding (encoding) is ancient and – like counting – a basic element of human communication. The representation of dual **numbers** is a dual **code**; a pure binary code. In Sects. 1.11, 1.12, 1.13, 1.14 and 1.15, learn the difference between dual numbers on the one hand and digital information and codes on the other.

1.1 Decimal Number System

All number systems are built up according to the same regularity. It is always a matter of expressing a very specific number (= "quantity") by a symbol, the "number".

Consider Fig. 1.1: In the example, sets of points are drawn. For each set (= number), there is a different number sign (numeral) that stands for that number. For zero piece there is numeral "0", for one piece there is numeral "1" etc. You can easily see that one cannot represent all arbitrary numbers in this way, because the infinitely many numbers would also require infinitely many sets of different characters.

Without special measures, one would therefore have to have as many different symbols as possible quantities, i.e. an infinite number! This is, of course, completely impossible. That is why the same symbols are used repeatedly. These basic symbols are the *numerals*.

© The Author(s), under exclusive license to Springer-Verlag GmbH, DE, part of
Springer Nature 2022
H.-J. Adam, M. Adam, *PLC Programming in Instruction list according to IEC 61131-3*, https://doi.org/10.1007/978-3-662-65254-1_1

Fig. 1.1 Sets of points:
Numbers stand as symbols or
numerals for the number
of points

Quantity:	Digit
	0
●	1
●●	2
●●●	3
●●●●	4
●●●●●	5
●●●●● ●	6
●●●●● ●●	7
●●●●● ●●●	8
●●●●● ●●●●	9

For larger quantities than nine pieces, new numerals are no longer used, but these numbers are formed by a composition of the familiar symbols 0 … 9. In the German and English language, however, there are even more number words: Zehn (ten, 10), Elf (eleven, 11) and Zwölf (twelve, 12). Only after that the new number words are formed by compositions of the basic number words: Thirteen (three – ten), Fourteen (four – ten) etc.

Exercise 1.1

Put together the number of different number words for different foreign languages you know! Check how the numbers between "10" and "20" are formed in different languages!

1.2   Bundling

If the set is larger than the supply of introduced number symbols, one must bundle. In all languages, the larger numbers are composed of the finitely many number words of a basic supply. In the tens or decimal system, one must always form bundles of ten each. The resulting number (quantity) of "bundles of ten" is again expressed by one of the number symbols. In Fig. 1.2 a quantity of 24 points is bundled.

The complete bundles of ten are named with the number of bundles, adding the syllable "-ty" to identify them as bundles of ten. (e.g., forty, sixty). Exceptions apply to twenty (instead of twoty), thirty (instead of threety) and fifty (instead of fivety). The number in the example from Fig. 1.2 is therefore called "Twenty-four".

Fig. 1.2 Bundle quantity of
24 pieces

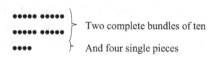

Fig. 1.3 Bundle for
two-hundred-thirty-seven

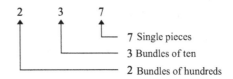

By bundling, one achieves that the complete bundles are created and the remaining numbers are always nine or smaller when bundling by tens. As a practical application, you can imagine that ten eggs are packed in an egg box. Ten of these boxes are packed in a carton, of which ten are again packed on a pallet. So the eggs, boxes, cartons or pallets are divided into bundles of 10. A box contains 10 eggs, a carton 100, a pallet again ten times as many, that is 1000 eggs. You can now divide any number of eggs among the pallets, cartons, boxes and remaining individual pieces. A maximum of nine of each container is required. In Fig. 1.3 you can see an example of bundling: For 237 eggs you need 2 cartons, 3 boxes and 7 eggs remain single.

Exercise 1.2

Bundle the number of strokes in the field "0" to bundles of ten in Fig. 1.4. For each complete bundle that you cross out in the right-hand field "0", write down a marker (flag) in the field "1" to the left as a mnemonic. If the field "1" contains more than ten strokes, the strokes in this field must also be bundled with bundles of ten! Draw for each 10 strokes in the field "1" in the left field "2" again one stroke. (This counting rule must be continued until there are no more than 9 flags in any field). Write down the number of remaining flags under the respective field.

1.3   The Decimal Position System

Each digit of a number corresponds to a certain quantity, which is calculated from the digit multiplied by the place value. The difference between the place values is always the factor 10, which is why such a number system is called "decadic" or "decimal".

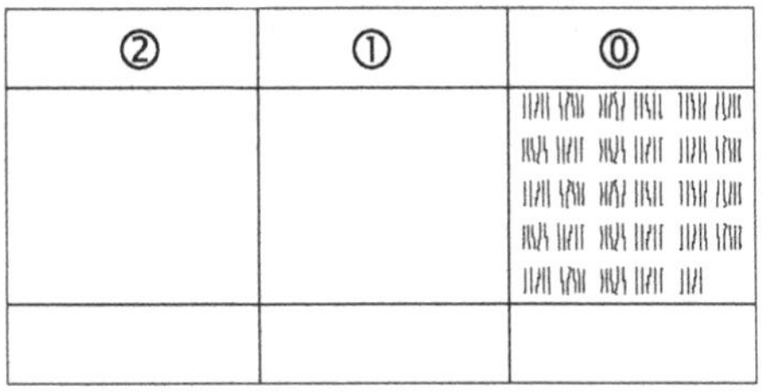

Fig. 1.4 Bundle fields for
Exercise 1.2

②	①	⓪
		⑪⑪ ⑪⑪

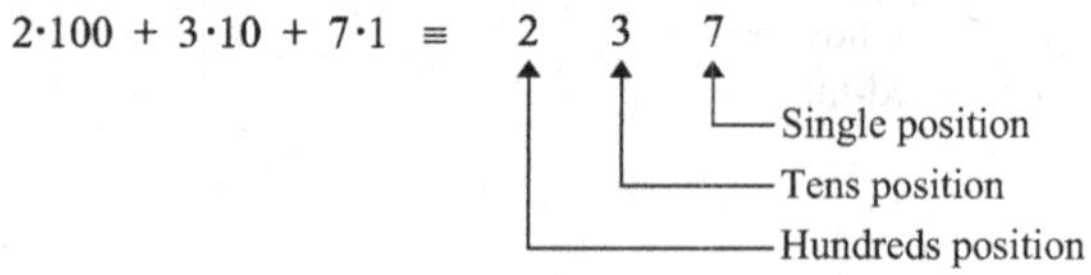

Fig. 1.5 Decimal number $2 \cdot 100 + 3 \cdot 10 + 7 \cdot 1 \equiv$ 2   3   7

                                                                      Single position
                                                                      Tens position
                                                                      Hundreds position

Our "numbers" represent the number of singles, tens, hundreds, thousands, etc. bundles! Because there are at most nine elements in each case, the ten different number words "zero" to "nine" are sufficient to represent all numbers. The number two hundred and thirty-seven is shown in Fig. 1.5.

Table 1.1 lists the values of the individual positions in the decimal system. To simplify the notation, we only note the respective numbers for the numbers, but not the factors 1, 10, 100, etc. (see Fig. 1.5). This procedure allows a compact representation of the numbers, but has some consequences:

Order The agreed order of the digits must be strictly adhered to. If you put the digits in a different order, the value of the number changes. Therefore, both the digit itself and the position of this digit within the number determines its numerical value. The number "one hundred twenty three" is written in digits as "123". If you swap the order of the digits, the numerical value changes significantly. This is different from the Roman and Egyptian numbers! (see Sects. 1.4 and 1.5) Unfortunately, in the German language the order of the number words for the tens and ones digits are swapped: "ein-hundert-drei-und-zwanzig" instead of "einhundert-zwanzig-drei". Because we have become accustomed to this from an early age, we do not notice it. In other languages it is often different, for example in English: "one-hundred-twenty-three".

Numeral The respective number signs for the numbers may only be single digits. This is the case with the "normal", decimal number system and therefore does not cause us any problems so far. With other systems, e.g. with hexadecimal numbers (Sect. 1.10), new number signs must be "invented" for this reason.

Table 1.1 Decimal system

...	Position 5	Position 4	Position 3 Thousands	Position 2 Hundreds	Position 1 Tens	Position 0 Ones
...	100,000	10,000	1000	100	10	1
...	10^5	10^4	10^3	10^2	10^1	10^0

Zero The numeral 0 is of special importance, without which a correct position representation is not possible. If no dashes remain in a field, this position must be occupied by the zero. Neither with the Roman nor with the Egyptian numbers is the zero necessary!

Exercise 1.3

In Fig. 1.4 of Exercise 1.2, write above each box the value that each of the dashes in that box represents. Write down the total number not only in numbers, but also in words!

1.4   Roman Numerals

As an example of a non-decimal number system, we would like to cite the Roman numeral symbols. Here the number symbols I, V, X, L, C, D and M mean the values 1, 5, 10, 50, 100, 500 and 1000. Larger numbers are expressed by repeating the symbols, subtracting or adding the values.

The number four is represented by IV ($-1 + 5$), six by VI ($5 + 1$), the number nine by IX ($-1 + 10$), 1998 by MCMXCVIII ($1000–100 + 1000–10 + 100 + 5 + 1 + 1 + 1$) and 2012 by MMXII ($1000 + 1000 + 10 + 1 + 1$).

We will not go into this number system in detail here. For further information we refer to the relevant literature.

1.5   Egyptian Numbers

The ancient Egyptian number system seems remarkable to us. The ancient Egyptians had a number system (Fig. 1.6), which is similar to our decimal system at first sight, because the basic digits differ by powers of ten. Each basic digit has its own symbol. The order in

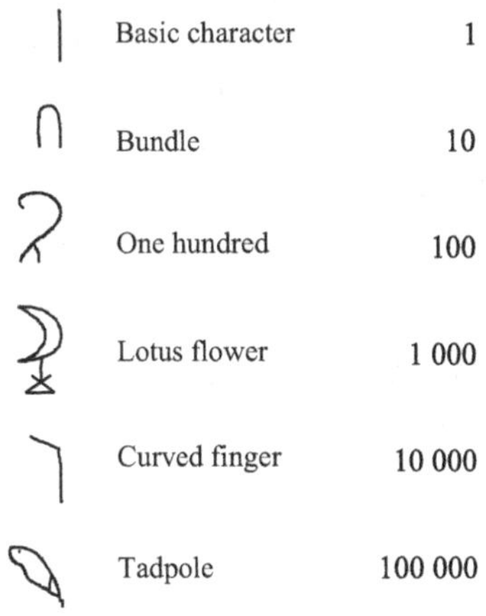

Fig. 1.6 Egyptian number system (hieroglyphs)

	Basic character	1
	Bundle	10
	One hundred	100
	Lotus flower	1 000
	Curved finger	10 000
	Tadpole	100 000

Fig. 1.7 Example of an
Egyptian hieroglyph

    2   3   2   4   1   3

Fig. 1.8 For Exercise 1.4

a

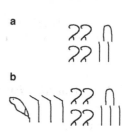

b

which the various symbols are displayed is therefore irrelevant. The numerical value is not
bound to the position. The symbols would therefore not need to be sorted as in Fig. 1.7,
but could be arranged arbitrarily in the order.

Exercise 1.4

What is the value of the hieroglyphs in Fig. 1.8?

1.6   Binary System, Dual Number System

Bundling, as in the decimal system, can be done with any basic number. In earlier times,
the division into 12 pieces ('dozen') and 12 dozen ('gros') was common. The smallest
basic number with which bundling is possible is two. This system of twos or dual system
is based on a bundle of twos. Figure 1.9 shows how dual numbers are produced by splitting
them into bundles of two. The dual numbers are also called binary numbers.

The values of the individual positions in the dual system are summarized in Table 1.2.

The dual number 101011 is the number 43 in the decimal system (Fig. 1.10). This number representation is somewhat unwieldy for us humans because the numbers quickly
become very long. For machines, however, there are great advantages.

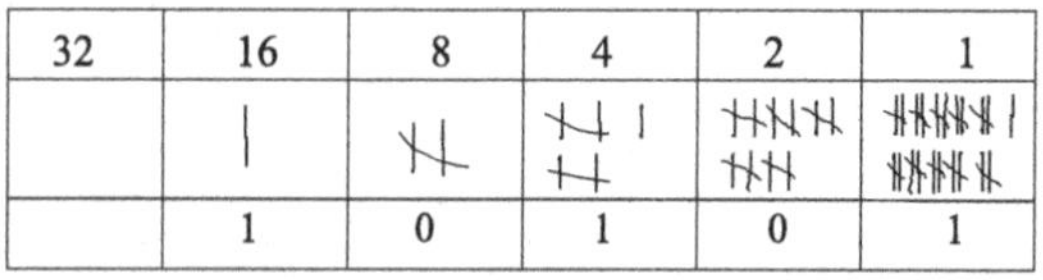

Fig. 1.9 Example of bundling
21 pieces as a two-
piece bundle

32	16	8	4	2	1
	ǀ	卅	卅 ǀ 卅	卅卅卅 卅卅	卅卅卅 ǀ 卅卅卅 卅
	1	0	1	0	1

Table 1.2 Dual system

	Position 5	Position 4	Position 3	Position 2	Position 1	Position 0
	32	16	8	4	2	1
	2^5	2^4	2^3	2^2	2^1	2^0
...	Position 11	Position 10	Position 9	Position 8	Position 7	Position 6
...	2048	1024	512	256	128	64
...	2^{11}	2^{10}	2^9	2^8	2^7	2^6

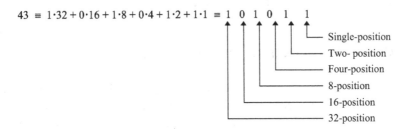

$$43 \equiv 1\cdot32 + 0\cdot16 + 1\cdot8 + 0\cdot4 + 1\cdot2 + 1\cdot1 \equiv 1\;0\;1\;0\;1\;1$$

- Single-position
- Two- position
- Four-position
- 8-position
- 16-position
- 32-position

Fig. 1.10 Decimal number and dual number: for example 43_{dec} and 101011_{dual}

Fig. 1.11 Bundling of two for Exercise 1.5

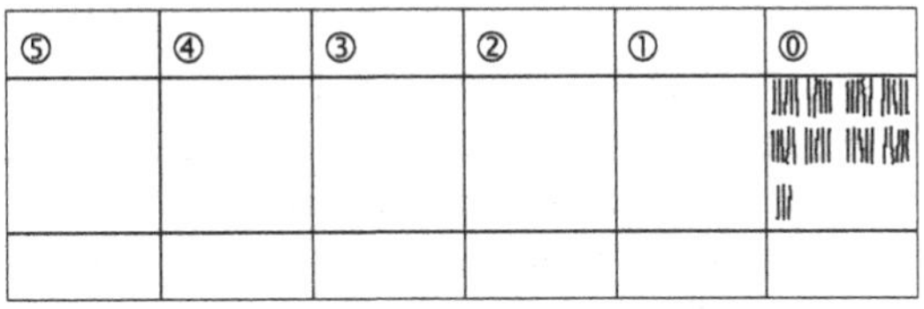

⑤	④	③	②	①	⓪
					IIII IIII IIII IIII IIII IIII IIII IIII III

Exercise 1.5

Carry out a bundling of two in Fig. 1.11. For every two flags, a flag must be entered in the field to the left.

Exercise 1.6

Convert the dual numbers to decimals:

a.	011 =	b.	10101010 =	c.	111 =	d.	10010010 =
e.	100 =	f.	01010101 =	g.	10000 =	h.	10100 =
i.	110 =	j.	10000000 =	k.	1111 =	l.	10000101 =

1.7   Computers Work with Dual Numbers

Computers do not know numbers, only voltages (volts) and currents (amperes). One could assign the numbers to certain voltage values, e.g. 1 volt each. The number 'two' would then correspond to 2 volts, the number 'thirteen' to 13 volts, and so on. Technically, however, it is not possible to use voltages that are too large, nor is it possible to represent the voltages sufficiently accurately in arbitrarily small steps. Computers based on such an 'analogue' number system do not achieve very great accuracy. Even a decimal system would still require ten different voltage values for each decimal place, which would not be technically easy to handle either.

To avoid such problems, one must use number systems that make do with fewer different number symbols. The system with the smallest number of different symbols is the two-symbol system or dual system. The two different symbols 0 and 1 can then simply be expressed by "voltage present" or "voltage absent".

1.8   Conversion of Decimal Numbers into Dual Numbers

Decimal numbers can be converted into dual numbers by bundling of two. However, one can also determine the power-of-two values contained in the decimal number (starting with the highest), determine the remainder value and here again determine the highest power-of-two contained.

Example 1.1 (Convert Decimal Number to Dual Number)

The decimal number 21 is to be converted into a dual number: The highest power of two contained is $16 = 1 * 2^4$. In the remainder $21-16 = 5$, the highest power of two is $4 = 1 * 2^2$. Now the remainder is: $5-4 = 1 = 1 * 2^0$. The two powers of two 2^3 and 2^1 do not occur, i.e. we have the values $0 * 2^3$ and $0 * 2^1$. So the dual number is: $\mathbf{1}*2^4 + \mathbf{0}*2^3 + \mathbf{1}*2^2 + \mathbf{0}*2^1 + \mathbf{1}*2^0$ Only the digits are written: $21_{dec} = 10101_{dual}$. ◀

Exercise 1.7

Calculate the corresponding dual numbers for the following numbers of the decimal system:

a.	3 =	b.	9 =	c.	15 =	d.	20 =
e.	31 =	f.	45 =	g.	99 =	h.	189 =
i.	200 =	j.	266 =	k.	278 =	l.	311 =
m.	499 =	n.	556 =				

1.9   Other Number Bases, Hexadecimal Numbers (Base 16)

Except for bundling with 10 or 2, bundling can be on any basis. An example may be the packing of 12 eggs in a box. The number 12 is also called 1 dozen. If again 12 dozen are packed in one carton each, then $12 * 12 = 144$ pieces are obtained. This number is a gross. This can be continued like this: 12 cartons on a pallet, 12 pallets in a truck, etc. In this case 12 is the number base.

In computer technology, one likes to use the base 16, i.e. 2^4. This base is called *hexadecimal* or *sedecimal* base. Because the base 2 is contained in the power of 4, there is a particularly simple connection between the dual system and the hexadecimal system, which makes the conversion of a number from one system to the other very simple. We will look at this in more detail in Sect. 1.10.

The number signs of the hexadecimal system for the values starting from 'Ten' must be one-digit, so that an unambiguous assignment to the positions is possible. The decimal number '26' is in hexadecimal form: $1*16 + 10*1$. Since only the coefficients (here '1' and '10') are written in the position system, the number '110' would result, which is not correct ($110_{hex} = 1*16^2 + 1*16 + 0 = 272_{dec}$). The number 'ten' must be written in one digit as 'A', so that the correct HEX-number is called for $26_{dec} = 1A_{hex}$.

> Since the hexadecimal system requires 16 different (one-digit!) number signs, the first letters of the alphabet are used from 'Ten' to 'Fifteen'.

The number signs (digits) in the hexadecimal system are listed in Table 1.3. The (decimal) valences of the digits of a HEX number are given in Table 1.4.

Table 1.3 HEX digits

HEX digit:	0	1	2	3	4	5	6	7	8	9	A	B	C	D	E	F
Decimal:	0	1	2	3	4	5	6	7	8	9	10	11	12	13	14	15

Table 1.4 Hexadecimal system or sedecimal system

...	Position 5	Position 4	Position 3	Position 2	Position 1	Position 0
...	10,048,576	65,536	4096	256	16	1
...	16^5	16^4	16^3	16^2	16^1	16^0

Example 1.2 (HEX Number and Decimal Number)

The number 333_{hex} means:

$$3*16^2 + 3*16^1 + 3*16^0 = 819_{dez}.$$ ◀

Convert the following hexadecimal numbers to decimal numbers:

a.	110 =	b.	CD =	c.	1234 =	d.	EF6A =
e.	3456 =	f.	10F2 =	g.	109 =	h.	AE29 =
i.	87E4 =	j.	6E5B =	k.	AFFE =		

Convert the following decimal numbers to hexadecimal numbers:

a.	9 =	b.	45 =	c.	99 =	d.	580 =
e.	910 =	f.	3030 =	g.	6550 =	h.	8750 =
i.	11111 =	j.	60000 =				

1.10   Hexadecimal Numbers and Dual Numbers

Dual numbers (binary numbers) can be converted particularly easily into hexadecimal numbers, because in the base 16 the base 2 occurs as power 2^4. For every 4 binary digits, one hexadecimal digit is therefore obtained.

Example 1.3 (Binary Numbers (Dual Numbers) and Hexadecimal Number)

In Fig. 1.12 we see an example of converting the HEX number 9D3F to the dual number 1001110100111111. ◀

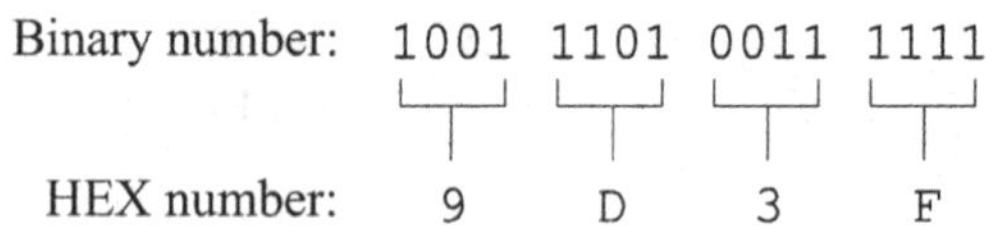

Fig. 1.12 Binary number and hexadecimal number

Binary number: 1001 1101 0011 1111

HEX number: 9   D   3   F

Convert the binary numbers from Exercise 1.6 into hexadecimal numbers. Then convert the HEX numbers into decimal numbers and check the result with your earlier solution.

1.11 Binary Code

The decimal number 837 as a dual number is a pure binary code, the valences of the individual digits correspond to the powers of two: $837_{dec} = 1101000101_{dual}$. For further examples, see Exercise 1.6.

1.12 BCD Code

The coding of decimal numbers into dual numbers can be carried out more clearly if no pure binary code is used according to the powers of 2, but the decimal digits are coded digit by digit. This results in the BCD code. The abbreviation stands for Binary Coded Decimals.

Example 1.4 (Relationship Between Decimal Number, Dual Number and BCD Number)

The number $837_{dec} = 1101000101_{dual}$ looks like this as a BCD number: 100000110111_{BCD}.

Decimal number:	8	3	7
BCD-coded number:	1000	0011	0111_{BCD}

◄

In contrast to the pure binary code representation with 10 digits, however, 12 digits are now necessary! The disadvantage that BCD numbers are longer than the corresponding pure binary ones is compensated by the better readability for humans. BCD representation is also more convenient for applications that need to display counter readings on a numeric display. However, for the computer it often means more complicated arithmetic operations – but we don't mind that.

Convert the numbers from the previous exercises into hexadecimal numbers and into BCD-coded dual numbers!

Table 1.5 Examples for the ASCII code

Characters	Bit sequence	Dec	Hex	Characters	Bit sequence	Dec	Hex
A	0100 0001	65	41	**0**	0011 0000	48	30
B			42	**1**			31
C	0100 0011			**2**	0011 0010		
a		97		**8**			38
b	0110 0010			**9**		57	
c			63	**(**	0010 1000		
.		46		**)**			29

1.13   ASCII Code

The notation with only two different characters, "0" and "1", is also suitable for an encoding of arbitrary characters, letters and even actions (=commands). The position of the characters '0' or '1' gives then of course no statement about the (numerical) value, because they are not numbers!

An important code used to represent typewriter characters in binary form is the ASCII code. The abbreviation stands for American Standard Code for Information Interchange. This American Standard Code for Information Interchange is a code that consists of 7 digits, is therefore a 7-bit code. Therefore, 2^7, i.e. 128 characters can be represented. It is possible to encode all letters, digits and some special characters with this code.

When setting up the 0,1 bit sequences representing the character, the assignment of the bit sequences is in principle arbitrary and has nothing to do with the numbers formed from the same bit sequences.

To remember the respective binary representation of the ASCII characters, it is usual to read the code like a numerical code, i.e. to assign the valences of the powers of two to the individual digits, i.e. to interpret the (arbitrary) bit sequence as a "number". Therefore, the representation for "A" (0100 0001) can be read as like a dual number $41_{hex} = 65_{dec}$. You say: the code for A is 41_{hex} or 65_{dec}, although this is actually wrong. You now know that the characters of the ASCII code are not numbers but symbols, codes!

Exercise 1.12

Add the missing values in the ASCII Table 1.5.

1.14 Dual Code – Dual Number

A dual number is not the same as a dual code! Pay close attention to the difference between the two:

Dualcode In a dual code, such as the ASCII code, no specific numerical value is represented. The two different characters "0" and "1" are used, but they do not result in a value. The order of the characters in the code is "arbitrary". By agreement in a standard, the meaning of the individual code characters is fixed.

Dual Number For a dual number, the two number signs "0" and "1" are used. Each digit is to be multiplied by the respective valence of the power of 2; this gives the value of the dual number.

1.15 Signals

Scientifically, a signal is the physical representation of a message. The signal transmits information from a transmitter to a receiver. The transmitter can be, for example, a thermocouple or a level transmitter. These signal transmitters are also called "sensors". The receiver is the computer or PLC automation device.

The signals enable an exchange of information. This is true for nature as well as for our technical world. Signals and signal transmission have always existed. Information exchange can take place over short distances through speech or gestures. Over longer distances, one uses smoke signals, drums, light signals, electrical impulses, etc.

Today, a computer that processes information, i.e. actually signals, can only deal with electrical signals. However, very few signals are originally in electrical form. In order for the computer to work with them, they must first be converted into electrical quantities such as voltage or current.

The electrical signals with which a computer works today have a binary character: they assume only two different states, namely 'current' or 'no current'; 'voltage' or 'no voltage'; '0' or '1'; 'low' or 'high', because such binary signals are particularly easy to realize technically. These binary signals can be generated, for example, by limit signal generators. The signals could then be: maximum level reached or not reached.

Logical Functions and Boolean Algebra

2

Abstract

In the previous chapter we described how the digital signals can be used to represent dual numbers or binary codes. The digital signals assume exactly two different states, which can be assigned either to the number set $(0,1)$ of the dual numbers or to the binary states $(0,1)$ of the codes.

But also logic gets along with two different states: a statement can be true, thus be correct (true), or it can not be true, thus be incorrect (false). The two logical states 'true' and 'false' can therefore also be expressed, like the dual numbers, by binary signals. Because of this fact, binary numbers and logic signals are often confused.

In Sects. 2.1, 2.2, 2.3, 2.4, 2.5, 2.6, 2.7, 2.8, 2.9, 2.10 and 2.11 you will learn about logical functions. You will see how the signals from sensors can be evaluated and linked. First of all, we will look at sensors that only give two different signals (e.g. 'On'/'Off', 'Open'/'Closed', 'Underrange'/'Overrange').

In Sects. 2.12, 2.13, 2.14, 2.15, 2.16, 2.17, 2.18 and 2.19 you will learn some possibilities how a logical function can be realized by logical circuit elements. This procedure is called "circuit synthesis". The behaviour of the process can be recorded in a logic table. From this table the combination of logical circuit elements can be derived. In most cases, however, this does not directly result in the simplest circuit, i.e. the circuit with the smallest possible number of switching elements. You will get a small insight into the methods of circuit simplification in Sects. 2.20, 2.21 and 2.22.

From the basic functions: Negation, AND- and OR-operation all logical operations can be composed. Some of these composite functions are very important in technology and are therefore used like basic functions.

H.-J. Adam, M. Adam, *PLC Programming in Instruction list according to IEC 61131-3*, https://doi.org/10.1007/978-3-662-65254-1_2

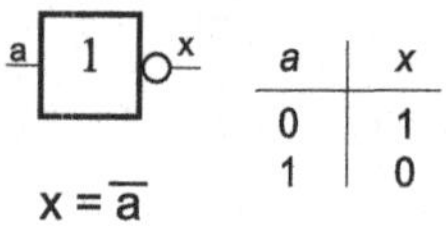

Fig. 2.1 Negation

$$a\boxed{1}\!\!\circ\!\!^{x}$$

a	x
0	1
1	0

$$x = \overline{a}$$

2.1   Negation (NOT Function)

> Function equation : $X = \neg a$ oder $X = \overline{a}$

This function always outputs the opposite of the input variable (Fig. 2.1). The output variable X always has the opposite value to the input variable a. As an example, you could imagine the sentence 'It is raining, then it is not dry'. The input variable 'rain' gives 'dry' as the output variable, but negated. The function equation marks the negation with an overstrike of the variable identifier. Say it like this: 'X is equal to a *not*', so the overstrike is spoken as an appended 'not'.

2.2   Identity (EQUAL Function)

> Function equation : $X = a$

The output variable X has the same value as the input variable a (Fig. 2.2). At first, this seems to be senseless; however, it often happens that two negations occur one after the other for technical reasons, which then "cancel" each other out and thus represent the equal function altogether. Sometimes you need an amplifier for a too weak logical signal. If this amplifier is done as an equal-function, the logical signal is not changed but only amplified.

Exercise 2.1 (EQUAL21)[1]

Form the equal function from two non-functions!

2.3   Conjunction (AND Function)

> Function equation : $X = a \wedge b$

The output variable is '1' if all input variables are '1' (Fig. 2.3). The AND operation states that an event only occurs if all conditions are fulfilled at the same time. Linguistically, you can express it like this, "If it rains AND I go outside, I'll take the umbrella."

---

[1] If you have a digital trainer (construction kit) available, you can perform the tasks with it. Instead of a hardware kit, you can also use a simulation program for the PC. For many exercises, you can find suggested solutions on the authors' website under the name given in brackets after the exercise number.

Fig. 2.2 Identity

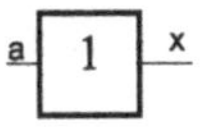

$$x = a$$

a	x
0	0
1	1

Fig. 2.3 Conjunction

$$x = a \wedge b$$

a	b	x
0	0	0
1	0	0
0	1	0
1	1	1

In the function table for "AND" there is only one '1' at the output. *Note:* The symbol "∧" for "AND" can be read as a simplified 'A'.

Exercise 2.2 (AND21)

Check the basic AND operation with the digital trainer. Make the assignments for the state of the lamp: $0 =$ 'off' and $1 =$ 'on'.

2.4   Heating Control (Two-Point Control)

To keep the room temperature even, you can proceed as follows: switch off the heating whenever it is warm enough, and switch it on again when the temperature falls below the desired level. This is called two-point control. This control is very easy to realize; therefore it is used very often. A bimetal switch can be used as a sensor and actuator, which opens a contact when a maximum value is exceeded and closes the contact again when the temperature falls below a minimum value, thus switching on the heating. In this way, the temperature can be kept almost constant.

The two states of the sensor (switch open/closed) served me here however first only for the description of the principle. In reality, of course, the sensors must generate signals that correspond to the conditions for '1' or '0'. So when used in a logic circuit, the sensor always gives a logic signal.

Exercise 2.3 (HEAT21).[2]

A reaction vessel is to be kept at a constant temperature with the heating switched on via a contact thermometer as a limit signal generator. Develop the function table, the logic diagram and the function equation. Check the result with the digital trainer.

---

[2] The drawing associated with this exercise (Fig. 2.4) is not complete. It is part of your "homework" to complete the drawing for this and many other exercises. (For solutions, see the authors' website).

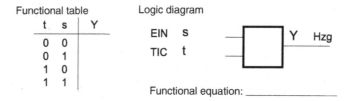

Fig. 2.4 For Exercise 2.3

```
Use the following assignments:
S = 1: ON Heating main switch switched on
t = 1: TIC temperature not reached
Y = 1: Hzg heating up
```

Exercise 2.4 (MIXER21)

Let a LED indicate a "Process Ok" signal ($x = 1$) when a stirrer is running ($a = 1$) and the level is reached ($b = 1$). See Fig. 2.4 for function table and logic diagram.

2.5   Negation of the Input

$$\boxed{\text{Function equation}: y = a \wedge \bar{b}}$$

Exercise 2.5 (ALARM21)

Create a function table, logic diagram and function equation for the following problem: An alarm ($Y = 1$) is to be triggered when a stirrer is switched on ($a = 1$) and the fill level has not yet been reached! ($b = 1$: level reached). In this task, the "opposite" is applied to input b as in the previous Exercise 2.4.

To solve, you can first set up the function table (see Fig. 2.5). Also add a column with 'b not' ($\bar{b}$); then you can form the result column for y quite easily. In the result column, 'zero' appears three times and 'one' only once. Thus, the problem can be solved with an AND term. Test your result with the digital trainer!

2.6   More Than Two Input Variables

$$\boxed{\text{Function equation}: X = a \wedge b \wedge c}$$

Fig. 2.5 for Exercise 2.5

a	b	b̄	y
0	0	1	0
1	0	1	1
0	1	0	0
1	1	0	0

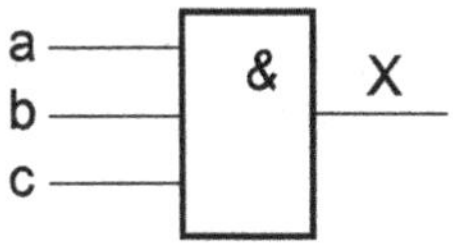

Fig. 2.6 Three inputs

Fig. 2.7 Function table with
three variables

a	b	c	Y
0	0	0	0
1	0	0	0
0	1	0	0
1	1	0	0
0	0	1	0
1	0	1	0
0	1	1	0
1	1	1	1

If only two input variables a and b occur, there are a total of four different combinations. You can combine these four combinations with a third input c (Fig. 2.6). If you first select this input as '0' and then as '1', you will get four combinations each (Fig. 2.7).

With three variables, you get eight different possible combinations, twice as many as with two inputs. The function table therefore consists of eight lines. With each additional variable, the number of combinations is doubled again.

2.7   AND Operation as Data Switch

With the help of the AND operation you can realize a "data switch". That means you get the possibility to switch on and off another logical signal with a logical signal. Among other things, a sensible application for this is the switching of a periodic signal (flashing signal). The second signal can then be used to switch a light emitting diode to "flashing light" or to "off".

Now how do you build such a switch? Look at the function table for the AND element and compare the output x with the input b (Fig. 2.8). First concentrate on the two lines where $a = 0$. The output x is always 'zero', the input b is thus "blocked" (Fig. 2.9).

Now look at the two lines with $a = 1$. Can you see that in this case it is always $x = b$; the input b is in a sense 'connected through' directly to the output x? In the other case, the output x is always '0'.

The AND element can serve as a switch for digital signals: The "switch" is the input a, the signal to be switched is applied to b. When *switched off*, the output is "0".

Fig. 2.8 AND operation as
data switch

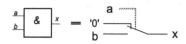

a	b	x
0	0	0
0	1	0
1	0	0
1	1	1

Fig. 2.9 Circuit diagram
"Data switch"

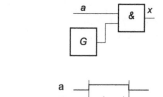

Fig. 2.10 "Flasher": the
generator signal G can be
switched on and off at input a

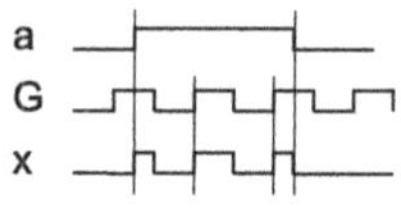

Fig. 2.11 Example of signal
curve for a turn signal
according to the circuit in
Fig. 2.10

Example 2.1 (Switch Flashing Signal)

In order to switch a LED as a flashing signal, one needs a generator for the periodic signal and an on/off signal which is applied to the input a of the AND element. The circuit is shown in Fig. 2.10. The light emitting diode is connected to the output x. The time course of the individual signals can thus be shown: The signal from the generator reaches output x only as long as the signal $a = 1$. Figure 2.11 shows an exemplary course of the output signal: The pulses from the generator "get through" as long as the input $a = 1$. ◀

Exercise 2.6 (FLASH21)

Use the digital trainer to build a circuit where you can use a switch to turn the flashing signal on and off.

2.8   Disjunction (OR Function)

$$\boxed{\text{Function equation}: X = a \vee b}$$

This function links two (or more) input variables. The output variable is always 1 if one or more input variables are 1. An event should therefore already occur if at least one of the

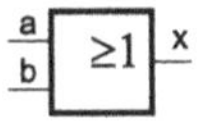

Fig. 2.12 Disjunction

$$x = a \vee b$$

a	b	x
0	0	0
1	0	1
0	1	1
1	1	1

Fig. 2.13 Function table for
OR with three inputs

a	b	c	Y
0	0	0	0
1	0	0	1
0	1	0	1
1	1	0	1
0	0	1	1
1	0	1	1
0	1	1	1
1	1	1	1

conditions is fulfilled. Linguistically, this corresponds to the sentence: "If it rains OR snows, I take the umbrella." (Fig. 2.12).

> In the function table for "OR" there is only one '0' at the output. In the function table for "OR", everything is one at the output except in one line.

You could also express the relationship between outputs and inputs the other way round: The output x is 'zero' if both inputs a and b are 'zero'. But beware, if you see it this way, you might confuse the two links AND and OR. To avoid this danger of confusion, it is usual to consider only those cases in which the output x is equal to 'one'.

More Than Two Input Variables

Function equation : $Y = a \vee b \vee c$

Consider Fig. 2.13. As always with the OR function, there is only a single zero in the result column.

2.9   OR Operation as Data Switch

Now look at the function table for the OR operation to see how the output is controlled by the variable a. Again, concentrate on the two lines with $a = 0$ or $a = 1$! In the "not switched" state the output is '1'. The OR element can serve as a switch for digital signals. In the *switched off* state the output is '1' (Fig. 2.14).

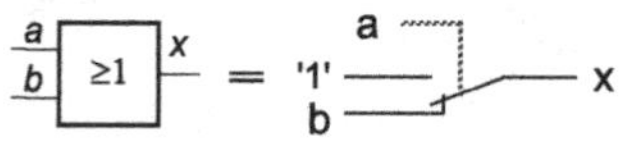

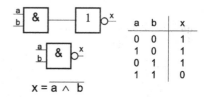

Fig. 2.14 Data switch implemented with OR operation

Fig. 2.15 NAND

a	b	x
0	0	1
1	0	1
0	1	1
1	1	0

$x = \overline{a \wedge b}$

Exercise 2.7 (FLASH22)

Now build a flashing circuit with on/off switching capability using an OR element. Draw the time curve of the signals. Compare the signals with those from Sect. 2.7. Use the function tables to show the different effects on the output signal x when the AND or OR operation is used as a data switch.

2.10 NAND Function (Negation of the AND Function)

$$\text{Function equation}: X = \overline{a \wedge b}$$

If you invert the output of an AND operation by a negator, you get the NOT-AND = NAND operation (Fig. 2.15).

Exercise 2.8 (NAND21)

Form a NAND function from an AND and a NOT function using the digital trainer and check the function table.

2.11 NOR Function (Negation of the OR Function)

$$\text{Function equation}: X = \overline{a \vee b}$$

If you invert the output of an OR operation by a negator, you get the NOT-OR = NOR operation (Fig. 2.16).

Exercise 2.9 (NOR21)

Form a NOR function from an OR and a NOT function with the digital trainer and check the function table.

Fig. 2.16 NOR

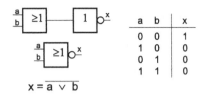

x = a ∨ b

For each of the following function tables, create the logic diagram and the function equation, or add the missing parts. Check the result each time with the aid of the digital trainer!

Function tables, logic diagrams and function equations for Exercise 2.10:

Function 1		Function 2	

Function table 1:   Logic Plan: Function table 2:   Logic Plan:

a	b	x
0	0	
1	0	
0	1	
1	1	

a	b	x
0	0	0
1	0	1
0	1	0
1	1	0

Function equation: $x = a \wedge b$ Functional equation:

Function 3		Function 4	

Function Table 3:   Logic Plan: Function Table 4:   Logic Plan:

a	b	x
0	0	
1	0	
0	1	
1	1	

a	b	x
0	0	
1	0	
0	1	
1	1	

Functional equation: Functional equation: $x = \overline{a} \wedge \overline{b}$

Function 5		Function 6	

Function Table 5:   Logic Plan: Function table 6:   Logic Plan:

a	b	x
0	0	
1	0	
0	1	
1	1	

a	b	x
0	0	
1	0	
0	1	
1	1	

Functional equation: Functional equation: $x = \overline{a} \vee b$

Function 7			Function 8		

Function Table 7:		Logic Plan:	Function Table 8:		Logic Plan:

a	b	x						a	b	x			
0	0	1
1	0	1
0	1	0
1	1	1

a	b	x
0	0	0
1	0	0
0	1	0
1	1	1

Functional equation: Functional equation:

Exercise 2.11 (LOGIC22)

Compare the result from Exercise 2.5 with the function table 2 from Exercise 2.10. Formulate similar examples of use for other functions from Exercise 2.10!

2.12   Creating a Function from the Function Table

In the previous exercises with the AND or OR elements, only *one* 'one' or only *one* 'zero' occurred in the output column. If more than one 'one' occurs in the result, then the function cannot be realized by a simple AND or OR circuit, but only by a combination of both. Now how can we determine this combination circuit? This question leads us to the notion of minterms or maxterms. In practice, the application of these terms is quite simple: You only need to decompose the total function into partial functions (namely the min- or maxterms), which are then combined.

Functions describe the relationship between output value and input value(s). In general, this relationship can be described symbolically by equations. The output value is usually written to the left of the equals sign, and the input values and their relationship are written to the right of the equals sign. For logical functions, the rules are described by "Boolean algebra". Here, only a very specific number of values and their combination can occur. Therefore, in the case of logical functions, the relationship between outputs and inputs can also be represented in the form of function tables.

Example 2.2 (Agitator with Switch-On Protection)

An agitator may only be switched on ($X = 1$) when the level is reached ($a = 1$) and the inlet valve is closed ($b = 0$) or the level is not reached and the valve is open.

In this task, you get a '1' twice in the output column for X. In order to obtain only one '1' (and thus to be able to apply the AND operation), you decompose the function X into the "subfunctions" X_1 and X_2, each of which contains only one '1' and can consequently be realized by AND operations of the two inputs. If necessary, one or both inputs must be negated. You have already practiced creating these functions for two inputs in detail in Exercise 2.10.

To obtain the complete function for X, the partial functions must be combined with an OR function in a second step (Fig. 2.17). ◄

a	b	X		a	b	X_2	X_1
0	0	0		0	0	0	0
1	0	1	$\equiv$	1	0	1	0
0	1	1		0	1	0	1
1	1	0		1	1	0	0

Fig. 2.17 Function table for the agitator circuit

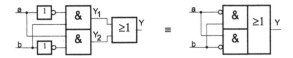

Fig. 2.18 Solution for Example 2.2

Fig. 2.19 EXOR

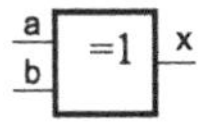

$$x = a \veebar b$$

a	b	x
0	0	0
1	0	1
0	1	1
1	1	0

For Example 2.2, the functional Eqs. (2.1), (2.2) and (2.3) are obtained in sequence, and the result is shown in Fig. 2.18.

$$X_1 = \left(\bar{a} \wedge b \right) \tag{2.1}$$

$$X_2 = \left(a \wedge \bar{b} \right) \tag{2.2}$$

$$X = X_1 \vee X_2 = \left(\bar{a} \wedge b \right) \vee \left(a \wedge \bar{b} \right) \tag{2.3}$$

> If there is more than one '1' in the result column of the function table, so many partial functions are formed that each contains only one '1'. The partial functions are linked with OR.

2.13 EXOR Function (Antivalence)

$$\boxed{\text{Function equation}: X = \left(\bar{a} \wedge b \right) \vee \left(a \wedge \bar{b} \right) = a \veebar b}$$

The function created in Example 2.2 is also known as EXOR. The name stands for Exclusive OR. It has its own switch character (Fig. 2.19). Note that with this function, unlike the "ordinary" OR, the output is 'one' only if either one or the other input is 'one'.

Fig. 2.20 Logic diagram
EXOR: display variants

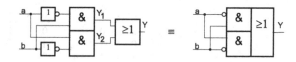

2.14 Disjunctive Normal Form (AND-Before-OR), Minterms

Normally the function table is always created first. We have already seen how you can develop the circuit from this in the example 'MIXER'. We got the EXOR or antivalence circuit (Figs. 2.18 and 2.20).

We had obtained the equation of the function by looking at the lines in which the result $X = 1$. These equations are called "minterms". They are created by ANDing the input values; if an input value is '0', the corresponding variable must be negated:

$$X_1 = \bar{a} \wedge b \qquad\qquad (2.4)$$

$$X_2 = a \wedge \bar{b} \qquad\qquad (2.5)$$

These two partial equations are then linked with OR. So we get an equation in the form AND-before-OR:

$$X = X_1 \vee X_2 \qquad\qquad (2.6)$$

$$X = (\bar{a} \wedge b) \vee (a \wedge \bar{b}) \qquad\qquad (2.7)$$

2.15 Brief Description of the Logic Diagram

For a clearer representation, the symbols of the individual switching elements can be drawn adjacent to each other. The negations can also be displayed directly at the input. The two circuits in Fig. 2.20 show both types of representation.

Exercise 2.12 (LOGIC23)

For each of the following functions 9 to 14, create the function table, the logic diagram and the function equation, or complete the missing parts. Check the result each time with the aid of the digital trainer!

Function 9 for exercise 2.12

Function Table 9:   Logic Plan:

a	b	Y	Y_1	Y_2
0	0	0	0	0
1	0	0	0	0
0	1	1	1	0
1	1	1	0	1

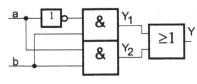

Functional equation:

Function 10 for exercise 2.12

Function Table 10:   Logic Plan:

a	b	Y	Y_1	Y_2
0	0	1		
1	0	1		
0	1	0		
1	1	0		

Functional equation:

Function 11 for exercise 2.12

Function Table 11:   Logic Plan:

a	b	Y	Y_1	Y_2
0	0			
1	0			
0	1			
1	1			

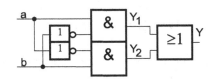

Functional equation:

Function 12 for exercise 2.12

Function Table 12:   Logic Plan:

a	b	Y	Y_1	Y_2
0	0			
1	0			
0	1			
1	1			

Functional equation: $Y = Y_1 \vee Y_2 = (\overline{a} \wedge \overline{b}) \vee (a \wedge b)$

Function 13 for exercise 2.12

Function table 13:   Logic Plan:

a	b	Y	Y_1	Y_2
0	0			
1	0			
0	1			
1	1			

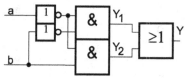

Functional equation:

Function 14 for exercise 2.12

Function table 14:   Logic Plan:

a	b	Y	Y_1	Y_2
0	0		0	0
1	0		1	0
0	1		0	0
1	1		0	1

Functional equation:

2.16 Two-Way Connection

You are certainly familiar with the application of a two-way connection: In a corridor, the light should be operable from two switches. Such a changeover circuit can be implemented with two two-way switches, each of which emits the logical signals "0" or "1". By changing the signal at one of the switches, the lamp is switched on or off.

Exercise 2.13 (SWITCH21)

Implement an alternating circuit with the aid of function elements. Create the function table, the logic diagram and the function equation. For realisation, assume that the lamp is off when both switches are in the '0' position. First complete the function table in Fig. 2.21.

```
Parts of the circuit:
Switch a,
Switch b,
Lamp Y
```

Fig. 2.21 Two-way connection for Exercise 2.13

a	b	Y
0	0	0
1	0	
0	1	
1	1	

Fig. 2.22 Circuit and function
table for Exercise 2.14

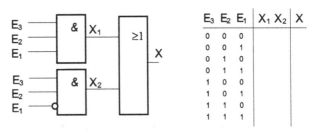

E_3	E_2	E_1	X_1	X_2	X
0	0	0			
0	0	1			
0	1	0			
0	1	1			
1	0	0			
1	0	1			
1	1	0			
1	1	1			

Functional equation: _____

2.17 Creating the Function Equation with More Than Two Inputs

The procedure described in Sect. 2.12 for generating the function equation also works in
the same way for function tables containing more than two input variables. Each partial
function contains only one single line with a '1'. The result is the OR-operation of these
minterms.

Exercise 2.14 (MINTERM21)

For the circuit shown in Fig. 2.22, first create the function table and then use the minterms to create the function equation.

2.18 Cross Connection

If two switching points are not sufficient for light switching, then a two-way circuit can be
extended with one or more "intermediate" cross over switches to a cross circuit so that a
lamp can be switched from three and more points. If only one switch is toggled, the lamp
is switched. If two switches are toggled at the same time, the state of the lamp does
not change.

Exercise 2.15 (SWITCH22)

Implement a cross-circuit for three switches using function elements. Create the function table, the logic diagram and the function equation (Fig. 2.23).
   Start again with the default: all switches in position '0' means lamp off. Starting
from this line, add further lines to the function table by searching for lines where only

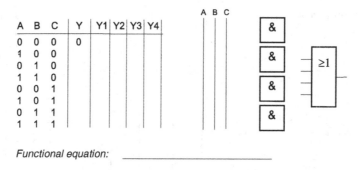

A	B	C	Y	Y1	Y2	Y3	Y4
0	0	0	0				
1	0	0					
0	1	0					
1	1	0					
0	0	1					
1	0	1					
0	1	1					
1	1	1					

Functional equation: _____

Fig. 2.23 Function table and circuit for Exercise 2.15

one switch state changes. Then the lamp state must also change. In this way, you can gradually fill in all the lines of the function table.

```
Parts of the circuit:
Switch A,
Switch B,
Switch C,
Lamp Y
```

2.19 Two- from Three-Sensors Logic

Very sensitive equipment may only be switched off in extreme emergencies, for example if an accident could occur. In other cases, safe monitoring must take place in any case, so that even faults in the monitoring equipment cannot have a negative influence. For both cases, the safety requirement can be met by using three different sensors for the same measured value. For example, if the exceeding of a critical temperature is to be monitored, three different thermometers would detect this value.

This safety circuit, in which a measured value or limit value is detected three times, is called a "two-out-of-three-sensors-logic". If *at least two* of these measuring systems signal that the limit value has been exceeded, there is a high probability that there is no faulty measurement. The safety device should become effective. This can be done, for example, by switching off the system. If only one of the sensors signals that the limit value has been exceeded, a faulty measurement *could* be present. This case should only be indicated by a warning device. In this way, false alarms can be avoided.

This type of safety circuit is particularly important for monitoring processes that are to run continuously and may only be switched off in extreme emergencies.

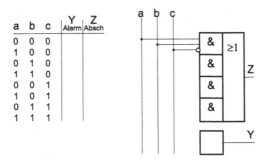

a	b	c	Y Alarm	Z Absch
0	0	0		
1	0	0		
0	1	0		
1	1	0		
0	0	1		
1	0	1		
0	1	1		
1	1	1		

Fig. 2.24 Safety circuit (two- from three-wire circuit): function table and circuit for Exercise 2.16

Exercise 2.16 (ALARM22)

Only if at least two of the three measuring systems indicate that the limit value has been exceeded, the safety shutdown and alarm should be triggered. If only one of the measuring systems indicates that the limit value has been exceeded, this "only" triggers the alarm.

Develop the function tables, logic diagrams, and functional equations for the safety shutdown and alarm (Fig. 2.24).

In the function table for the alarm Y there are many '1's and only one single row with a '0'. The standard procedure with the minterms as you have learned it so far works of course, but requires a huge effort, namely 7 AND-members with 3 inputs each and one OR-member with 7(!!) inputs.

If you don't want to drive this, you can consider by which simple function Y can be represented. Then only one switching element with 3 inputs is needed! Later we will learn a more entertaining method for this and similar cases. (If you are impatient, you can skip ahead to Sect. 2.22 for more information on 'Maxterms' and 'Conjunctive Normal Form').

Check the results found with the digital trainer. In addition, have a third output indicate when all three measuring systems report that the limit has been exceeded.

2.20 Transforming and Simplifying Functions

To realize a certain function, different circuits are always possible. These circuits which result in the same function table are called "identical circuits".

In the previous chapter, you determined the following functional equation in Exercise 2.10, table of Functions 4:

$$X_1 = \overline{a} \wedge \overline{b} \tag{2.8}$$

This function is identical to:

$$X_3 = \overline{a \vee b} \tag{2.9}$$

Fig. 2.25 Checking the equality of two functions using a table

a	b	x_1	x_2	x_3
0	0	1	0	1
1	0	0	1	0
0	1	0	1	0
1	1	0	1	0

Fig. 2.26 Function table for Exercise 2.17

a	b	c	Y_1	a∧b	a∧c	b∧c	Y_2
0	0	0	0				
1	0	0	0				
0	1	0	0				
1	1	0	1				
0	0	1	0				
1	0	1	1				
0	1	1	1				
1	1	1	1				

You can easily check this using the function table. Write in one column the results for X_1, in a second column the results for $X_2 = a \lor b$ and in the third column for X_3 (Fig. 2.25). You now only need to check that the two columns X_1 and X_3 are the same. (The table in Fig. 2.25 still has the column entered for $X_2 = a \lor b = \overline{X_3}$ clarity). Because the two columns X_1 and X_3 are equal, you can conclude that the functions $X_1 = \overline{a} \land \overline{b}$ and $X_3 = \overline{a \lor b}$ are also equal!

> To check the equality of two functions, you can compare the output columns in the respective function tables.

Exercise 2.17 (LOGIC24)

Check the equality of the two functions Y_1 (2.10) and Y_2 (2.11) by first determining the function tables in Fig. 2.26 and then comparing the corresponding columns:

$$Y_1 = \left(\overline{a} \land b \land c\right) \lor \left(a \land \overline{b} \land c\right) \lor \left(a \land b \land \overline{c}\right) \lor \left(a \land b \land c\right) \tag{2.10}$$

$$Y_2 = \left(a \land b\right) \lor \left(a \land c\right) \lor \left(b \land c\right) \tag{2.11}$$

Draw the logic diagram of the simplified function and plug it on the digital trainer! Check the results found with the digital trainer.

2.21 Boolean Algebra

The rules by which logical values can be calculated are described in Boolean algebra. These rules of logic were established by George Boole (1815–1864). Because Boolean algebra can be implemented in technology by means of switches, it is also called 'switching algebra'. The values that are linked together by the Boolean functions are called variables. Logic operations can be represented by tables, formulas (equations) or graphical symbols.

Knowledge of the rules of Boolean algebra enables you to "convert" more complicated functions into simpler ones. Similar to "normal" algebra, there are calculation rules for term conversion, so that the term Y_1 can be mathematically converted into the term Y_2 from the example.

Using rules of Boolean algebra, you can transform Eq. (2.12) into Eq. (2.13). You are to perform the equality of these two terms in Exercise 2.18 by comparing tables.

$$x = \left(\bar{a} \wedge b\right) \vee \left(a \wedge \bar{b}\right) \tag{2.12}$$

$$= \left(a \vee b\right) \wedge \left(\bar{a} \vee \bar{b}\right) = y \tag{2.13}$$

Because the "inner" parentheses must be calculated first, the representation of function x (Fig. 2.27) is called the *AND-before-OR* (the *disjunctive*) form, and function y is called the *OR-before-AND* (the *conjunctive*) form. We have already considered the former in Sect. 2.14. The conjunctive form is discussed in the following Sect. 2.22.

Exercise 2.18 (LOGIC25)

Complete the columns for x and y in the function table from Fig. 2.27. Then determine the equality. Draw the logic diagram of the simplified function and plug it on the digital trainer! Check the results found with the digital trainer.

a	b	$(\bar{a} \wedge b)$	$(a \wedge \bar{b})$	x	$(a \vee b)$	$(\bar{a} \vee \bar{b})$	y
0	0	0	0		0	1	
0	1	1	0		1	1	
1	0	0	1		1	1	
1	1	0	0		1	0	

Fig. 2.27 Function table for Exercise 2.18 and for (2.12), (2.13)

Fig. 2.28 Function table for	a	b	x
creating Eq. (2.14)	0	0	0
	1	0	1
	0	1	1
	1	1	1

Example 2.3 (Term Transformation)

Here is a second example of term transformation: Given the function table from Fig. 2.28. Using the procedure you learned at the beginning of this chapter, evaluate the terms in the table to arrive at the function:

$$x = \left(\bar{a} \wedge b\right) \vee \left(a \wedge \bar{b}\right) \vee \left(a \wedge b\right) \tag{2.14}$$

However, you already know that this function table represents the OR function, that is, it is expressed by $x = a \vee b$. Therefore, it must be true:

$$x = \left(\bar{a} \wedge b\right) \vee \left(a \wedge \bar{b}\right) \vee \left(a \wedge b\right) = \left(a \vee b\right) \tag{2.15}$$

◀

We would like to refrain from presenting the transformation of the terms in detail here; however, the transformation is possible by means of Boolean algebra. We limit ourselves here to proving the equality by comparing the two function tables, which you can certainly do on your own.

Exercise 2.19 (LOGIC26)

Prove the correctness of Eq. (2.16) by setting up and comparing the function tables:

$$Y_{neu} = S \vee \left(\bar{R} \wedge Y_{alt}\right) = S \vee \overline{\left(R \vee \bar{Y}_{alt}\right)} \tag{2.16}$$

This is a preliminary exercise for "memory devices", which are examined in more detail in Sect. 3.2.

2.22 Conjunctive Normal Form (OR-Before-AND)

In Sect. 2.14 we saw that there is a "disjunctive normal form". There is also a "conjunctive normal form". This is created when you only look at the rows in the function table with a '0' in the result. To uniquely get a '0', the input variables must be ORed. Only if both are '0', the result is also '0'. These terms are called 'maxterms'. Thus, in Exercise 2.16, Y represents a maxterm.

Fig. 2.29 EXOR: representation in disjunctive normal form (AND-before-OR) by means of function table

a	b	Y	Y_1	Y_2
0	0	0	0	0
1	0	1	1	0
0	1	1	0	1
1	1	0	0	0

Fig. 2.30 EXOR: representation in conjunctive normal form (OR-before-AND) by means of function table

a	b	Y	Y_3	Y_4
0	0	0	0	1
1	0	1	1	1
0	1	1	1	1
1	1	0	1	0

The terms must be combined with AND to the total result. For the example in Fig. 2.28 the term $x = a \vee b$, results directly because there is only *one* '0' in the result column.

In this procedure, variables are thus first linked by OR, then the results are linked by AND. This is now called the OR-before-AND operation. The AND-before-OR can always be transformed into the OR-before-AND operation and vice versa.

Unfortunately, because we are not learning Boolean algebra in this course, we cannot use the switching algebra rules to prove the equality of the terms. However, you can prove the equality of the two terms by comparing the function tables.

To demonstrate the design procedure using the maxterms, we take the already familiar function of the EXOR element (Fig. 2.17). At that time we generated the disjunctive normal form, where the subfunctions Y_1 and Y_2 each contained only a single '1' (Fig. 2.29).

Now here, in the table Fig. 2.30, we have created the functions Y_3 and Y_4 so that they each contain only a single '0'.

Look in Exercise 2.10, there you had created the functions in the function tables 5 and 8:

$$Y_3 = a \vee b \qquad (2.17)$$

$$Y_4 = \overline{a} \vee \overline{b} \qquad (2.18)$$

Y is '1' whenever both Y_3 and Y_4 are '1'. Therefore, for Y:

$$Y = Y_3 \wedge Y_4 \qquad (2.19)$$

$$Y = (a \vee b) \wedge (\overline{a} \vee \overline{b}) \qquad (2.20)$$

You see, *first* the variables a and b in the brackets are linked with OR; only then the linkage results are linked with AND: the execution order is *'OR before AND'*! You have already proved the equality with the term (2.7) in Exercise 2.18.

Memory Elements

3

Abstract

What do you do if only very brief signals occur in a process or in a plant, or if the effect of a brief keystroke is to last over a longer period of time? Of course: logical signals, e.g. alarm signals, must be able to be stored!

Storing means that a circuit continuously delivers the logic signal at its output, even if the triggering pulse has already ended, i.e. the pulse must "write" a logic signal into a memory. In this chapter you will learn how such memory elements are constructed.

3.1   Flip-Flops and Static Memories

In contrast to the previous circuits, the same input signals will no longer always produce the same output signals. Depending on the history, the same input signals can produce different output signals. At first this sounds absurd and arbitrary! But of course the circuit does not work according to a random principle.

Since the past history must be taken into account, values from the past must be stored. So there must be switching elements which have a "memory", a storage possibility. Here are two examples: A short press on a start button switches on a system, which should of course continue to run even if the button has been released and the input signals are again present as before. Or in counting tasks, the counting impulse (which is the same each time) must cause the circuit to display a number that is 1 higher; the numbers, i.e. the outputs of the circuit, are different each time, despite the same input impulses. The new number to be displayed depends on the previously stored past.

H.-J. Adam, M. Adam, *PLC Programming in Instruction list according to IEC 61131-3*, https://doi.org/10.1007/978-3-662-65254-1_3

Save Binary Signals

For the storage of a binary digit (bit), a memory location is required which either holds the information logically '0' (low, L-level) or logically '1' (high, H-level). Because digital technology does not know any intermediate states, such memory locations must switch back and forth between the two permitted states as "lightning-like" as possible when switching. In DIN 40900, this memory is referred to as a bistable element. The terms bistable flip-flop and flip-flop (FF) are also commonly used.[1]

If plant sections are to be switched by pushbuttons, the button presses must be stored. A button press only lasts for a short time. Therefore, because the effect should last longer than the keystroke, you must store the signal. As seen, flip-flops can store signals. Thus, you can use a pushbutton to generate a set signal to set a flip-flop. The set flip-flop triggers the corresponding action: a valve is opened, a motor is running, or a lamp is on. To switch off the action, either another pushbutton switch or logic is provided that generates a reset signal.

We will learn about different types of binary memories. The differences lie in the different ways in which or at what time the information is stored.

3.2   The RS Flip-Flop

Consider Fig. 3.1. With this flip-flop, storing is done like this: a '1' signal at the R input resets the output, i.e. to '0'; a '1' at the S input sets the output to '1'. After setting or resetting, '0' must be applied to both inputs again. As long as both input signals a and b remain at '0', the output signal Y does not change; the flip-flop outputs '0' or '1', depending on which signal was previously stored. So this is the *memory position*.

> *Attention:*
> It is not allowed to give both inputs '1' signal at the same time.
> In this case the output may be undefined.

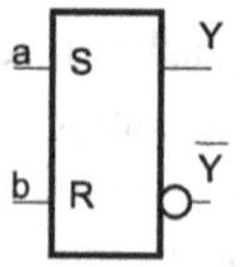

Fig. 3.1 RS bistable element

a	b	Y
0	0	Y (No change)
1	0	1 (Set)
0	1	0 (Reset)
1	1	?? (Forbidden)

---

[1]"Flip-Flop" is the American name. It is onomatopoeic and could be translated as "click-clack" in German.

Fig. 3.2 RS bistable element: function table with transition states

S	R	Y_{old}	$\bar{R} \wedge Y_{old}$	Y_{new}
0	0	0	0	0
1	0	0	0	1
0	1	0	0	0
1	1	0	0	-
0	0	1	1	1
1	0	1	1	1
0	1	1	0	0
1	1	1	0	-

Fig. 3.3 RS bistable element: feedback with delay

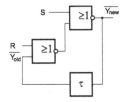

An RS flip-flop can be constructed with two NOR links. We can derive this with the following considerations: Because the new state for the output depends on the previous one, we introduce the state Y_{old} as the previous state and the Y_{new} new subsequent state in a function table in Fig. 3.2. In addition, we also entered $\bar{R} \wedge Y_{old}$ the column.

With Exercise 2.16 we have already somewhat prepared the problem shown here. If the forbidden cases are not considered, then '1' is always Y_{new} when $S = 1$ or is. $\bar{R} \wedge Y_{old} = 1$.

Therefore, for all permissible cases:

$$Y_{new} = S \vee \left(\bar{R} \wedge Y_{old} \right) \tag{3.1}$$

This term can be transformed as in Exercise 2.16 to:

$$Y_{new} = S \vee \left(\overline{R \vee \bar{Y}_{old}} \right) \tag{3.2}$$

$$\overline{Y_{new}} = S \vee \left(R \vee \overline{\bar{Y}_{old}} \right) \tag{3.3}$$

Now the question arises, how to get the 'old' state. Somehow the value of the output must be stored temporarily until it has been processed and the 'new' value has become stable. The simplest way to do this is to provide a delay (Fig. 3.3). The delay time τ must be long enough. Then the circuit already has the new value for Y at the input and the old value at the output for the time τ.

Fig. 3.4 RS bistable element: structure with standard components (**a**) as Fig. 3.3 without "feedback" (**b**) alternative representation

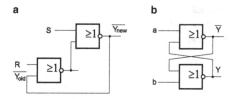

If one assumes that in practice the delay present in the gates anyway because of the signal propagation times is large enough, then the separate delay element can simply be omitted and replaced by direct "feedback" of the output to the input, as shown in Fig. 3.4.

Exercise 3.1 (RSFF31)[2]

Use the digital trainer to construct an RS-FF circuit from two NOR links and test the function. On which elements are the R and S inputs? Show that the outputs of the NOR gates are always opposite if a and b are not both '1'.

3.3   Alarm Circuit 1

An alarm circuit is used to monitor a process. The process emits an "alarm signal" in the event of a fault. The operating personnel must confirm the registration of the alarm by means of an acknowledgement signal. Because even a short alarm signal must not be overlooked, it must be stored. The acknowledgement signal resets the memory.

Exercise 3.2 (ALARM31)

Build the alarm circuit 1 (Fig. 3.5). The (momentary) signal 'Fault' should switch on the horn, which is switched off again by the acknowledgement signal.

Fig. 3.5 Alarm circuit 1 for Exercise 3.2

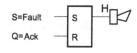

---

[2] Some of the tasks from this chapter will be performed later using the PLC. However, we recommend that you work on the tasks here and test them with a digital training device or simulation program if possible, even if you are eager to finally start programming PLCs. First, it's a preliminary exercise for programming because you can study the problems, procedures, and possible solutions using the more descriptive digital technology. And secondly, the graphical programming language "function block language" is practically identical to the digital technical representation.

Fig. 3.6 Preferred position
when switching on

Preferred situation
on power up:

Set at power up:

3.4   Defined Basic Position (Preferred Position)

Memory elements usually have an undefined state after the operating voltage is switched on, i.e. it is not possible to say whether the memory has a '1' or a '0' at the output after switching on. Often it is indifferent, but in many cases this can lead to malfunctions or even danger in practical applications.

To avoid this, special circuit measures can be taken to ensure that the memory element assumes a defined initial position after switching on (Fig. 3.6). This defined initial position is indicated in the symbol. In DIN EN 60617 (or DIN 40900), the initial basic position 0 is specified as $I = 0$. "I" stands for "Initial".

3.5   Priority of the Input Signals

The set and reset inputs of RS bistable elements must not be assigned a '1' signal at the same time, otherwise an undefined (forbidden) state will occur.

You can certainly think of many examples in practice where it is not guaranteed that the set and reset signals occur nicely one after the other. But what can you do if it cannot be ruled out that overlaps occur in the system? Yes, that's right, you have to create uniqueness by interlocking the signals. One of the two signals must be given priority, i.e. this priority signal blocks the other, lower priority signal.

RS Bistable Element with Set Priority
An AND circuit can be connected in front of the R input as a data switch (gate) (Fig. 3.7). Resetting is then only possible if no set signal, i.e. '0' signal, is present at input a. If a and b both carry a '1' signal at the same time, the reset signal at input b is interrupted and therefore ineffective. This means that only the set condition is fulfilled and the RS flip-flop is set (see Sect. 2.7).

Exercise 3.3 (RSFF32)

Build a set-priority RS bistable link from NOR gates and check the function.

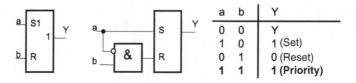

Fig. 3.7 Setting priority

RS Bistable Element with Reset Priority.

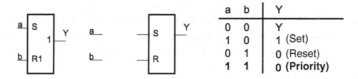

Fig. 3.8 Motor control: circuit and table for Exercise 3.4

Exercise 3.4 (RSFF33)

Add the logic diagram for an RS bistable element with priority of the reset signal in Fig. 3.8. Build the circuit with the digital trainer and check the function.

Exercise 3.5 (RSFF34)

Draw the logic diagram of an RS bistable element with priority of the first signal, i.e. the signal that is applied first.

3.6   Motor Control

For a motor, the two states "forward" or "clockwise" and "reverse" or "counterclockwise" must often be taken into account. There are therefore two different actions to be performed; you must therefore plan an RS flip-flop in your circuit for each of these states.

Design the respective control logic for both setting and resetting the flip-flop, i.e. you formulate the conditions that are to lead to the setting or resetting of the flip-flops of the corresponding state. These conditions can be: a '1' signal from a pushbutton or signals from the process.

Exercise 3.6 (LIFT31)

An elevator is to be controlled by briefly pressing pushbuttons UP and DOWN (Fig. 3.9). For this purpose, the motor is switched on in the direction of rotation clockwise and

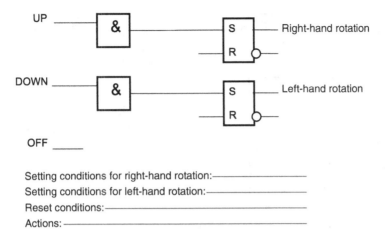

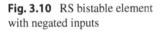

Setting conditions for right-hand rotation: ─────────────────

Setting conditions for left-hand rotation: ──────────────────

Reset conditions: ────────────────────────────

Actions: ──────────────────────────────

Fig. 3.9 Circuit and table for Exercise 3.6

Fig. 3.10 RS bistable element
with negated inputs

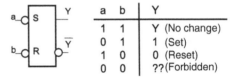

a	b	Y
1	1	Y (No change)
0	1	1 (Set)
1	0	0 (Reset)
0	0	?? (Forbidden)

anticlockwise. Switching off (immediate stop) is done by briefly pressing a third button. It must not be possible to switch on a direction of rotation if the opposite direction
is set, i.e. the direction of rotation can only be reversed after switching off, i.e. when the
lift has stopped. First formulate in words,

- how many memory elements (flip-flops) you need,
- by which signals the flip-flops are set,
- which signals reset the flip-flops,
- which actions must be performed when flip-flops are set.

Exercise 3.7 (LIFT32)

Extension of the previous task: In addition to the off switch, the drive must be switched
off and restarting blocked as long as the lubricating oil pressure detected by a pressure
switch is not present or the emergency stop switch is actuated. For this purpose, you can
also refer to the solution of Exercise 3.4.

3.7   Flip-Flop with Negated Inputs

For technical reasons, the setting and resetting of the flip-flops is often not performed with
'1' level, but with '0' level (Fig. 3.10). Compared to the 'normal' flip-flop, the $\overline{R}$ and $\overline{S}$ inputs
are negated. Thus, '0' is reset or set with the low level. This type is called "$\overline{RS}$-Flip-Flop".

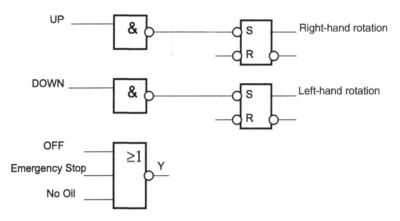

Fig. 3.11 Motor control: circuit and table for Exercise 3.8

Fig. 3.12 Write lock for
Example 3.1

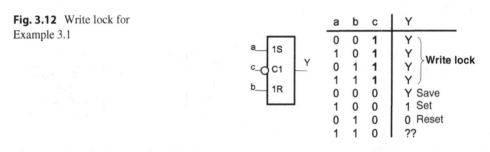

a	b	c	Y
0	0	1	Y
1	0	1	Y
0	1	1	Y
1	1	1	Y
0	0	0	Y Save
1	0	0	1 Set
0	1	0	0 Reset
1	1	0	??

(rows 1–4: Write lock)

Exercise 3.8 (LIFT33)

This slightly changes the solution of the previous task. To cancel the negations at the
inputs, you can use $\overline{\text{NAND}}$ or NOR elements. Complete the logic diagram for the motor
control, now using $\overline{RS}$ -flip-flops (Fig. 3.11)!

3.8   Clock-State Controlled Flip-Flops

Write Lock

It is often desired that the flip-flop may only be set or reset if a third signal is present. In
Fig. 3.12, this signal is present at the input $C1$. You can specify in the circuit design that
the set or reset signals may only take effect when a '0' is present at this control input. This
control signal is also called "clock signal". Setting or resetting is only possible if this clock
has a certain logical state, for example is in state '0'.

Fig. 3.13 RS bistable element
(flip-flop) with write lock for
Exercise 3.9

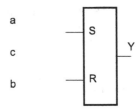

Example 3.1 (Write Lock)

A motor is to be switched on or off by one pushbutton each, but only if an enable signal
'0' is given to the control input by a main switch. In case of a '1' at the control input,
the old state should be maintained (Fig. 3.12).

Solution:

The memory contents can only be overwritten if a '0' signal is present at the control
input $C1$, i.e. Set and reset inputs of the memory are effective. If the control input carries '1' signal, the memory contents cannot be changed (write protection). To set up the
write inhibit, use a data switch ("gate circuit") in each of the supply lines to the S or R
input, consisting of an AND element (see Sect. 2.7). ◀

Exercise 3.9 (RSFF35)

Complete the circuit according to Fig. 3.13 so that you obtain the logic diagram for an
RS flip-flop with controlled input (write block). A '1' at the control input c is to block
the flip-flop.

Exercise 3.10 (RSFF36)

Design a RS flip-flop with write inhibit and additional set priority!

Flip-Flop with only One Data Input (Clock Input) and Write Inhibit

---

Exercise 3.11 (RSFF37)

You operate a circuit as shown in Fig. 3.14. In the simulator, you provide a switch for
input a (data input) and a pushbutton for C (write inhibit). Complete the time diagram
from Fig. 3.15 by drawing in the curve for signal Y and explain in your own words how
the circuit works!

Fig. 3.14 This circuit of a
flip-flop has only a single
data input a

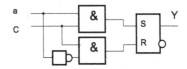

Fig. 3.15 RS bistable element
with write lock for
Exercise 3.11

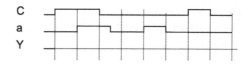

Read Lock

With the write inhibit just discussed, you prevent the flip-flop state from changing as long as control input C is blocked. The flip-flop is thus completely inactive during this time. Now, however, you want the new state to be accepted, but not to take effect at the output until later. The time at which the output indicates the new state is to be determined by a control signal. To solve this problem, you can simply interrupt the output signal of the toggle element with a data switch. The data switch can be implemented in a familiar way using an AND element. The flip-flop can be set or reset, but the change is not passed on as long as the data switch is blocked at its output.

Example 3.2 (Read Lock)

A stirrer is to be switched on and off by two buttons. As soon as a stop button is pressed ('0' signal), the motor run should be interrupted. Even while the stop button is pressed, the on/off buttons (as "preselection switches") should remain effective. ◄

Exercise 3.12 (RSFF38)

Implement an RS bistable element with disable input. Complete the logic diagram and the function table in Fig. 3.16! If the inhibit input carries a '1' signal, it should be possible to tap the memory contents at output Z of the circuit. If the inhibit input carries a '0' signal, the stored information is not available at the output of the circuit (read protection).

Exercise 3.13 (RSFF39)

Combine the two circuits "write lock" (Exercise 3.9) and "read lock" (Exercise 3.12)! Try out when, depending on the control signal c, the flip-flop and the output Z change.

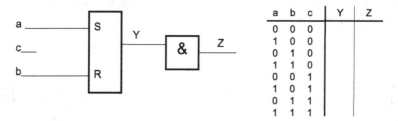

a	b	c	Y	Z
0	0	0		
1	0	0		
0	1	0		
1	1	0		
0	0	1		
1	0	1		
0	1	1		
1	1	1		

Fig. 3.16 RS bistable element with read lock for Exercise 3.12

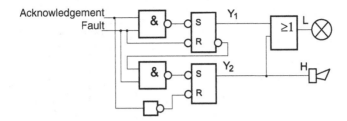

Fig. 3.17 Alarm circuit for Exercise 3.14

Fig. 3.18 Time diagram for
Exercise 3.14

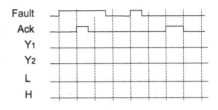

3.9 Alarm Circuit 2

Exercise 3.14 (ALARM32)

Analyse the function of the alarm circuit 2 Fig. 3.17. The alarm circuit is already much more refined than circuit 1 of Sect. 3.3. Understanding (analysing) the function of this circuit requires some concentration.

Investigate when which flip-flop is set or reset! When does the lamp L light up and when does the horn H sound?

First assume a basic state: both flip-flops are reset ($Y_1 = 0$ and $Y_2 = 0$), no signal is present (*Fault* = 0 and *Acknowledge* = 0). You can proceed in a particularly clear and systematic manner if you represent these time sequences in the diagram according to Fig. 3.18. To the right, enter the time segments, e.g. one segment per second.

Now let the signals come one after the other, first the disturbance *Fault*. Observe which of the flip-flops is set. Then give the acknowledgement signal *Ack*. Which flip-flops are set, which are reset?

Finally, try out how the circuit behaves if the fault lasts until the acknowledgement is received, or if it recedes beforehand.

3.10 Filling and Emptying a Measuring Vessel

This is a basic task of chemical process engineering: The components for the reaction are measured via one or more measuring vessels. The substances are then drained into the reaction vessel, where they are mixed, stirred, heated, etc., whatever the process requires.

Fig. 3.19 Measuring vessel

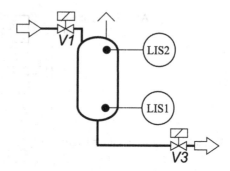

Exercise 3.15 (TANK31)

The measuring vessel in Fig. 3.19 is to be controlled in such a way that, after the "Fill" button has been pressed, water flows in via solenoid valve V1 until the "FULL" state is signalled by limit switch LIS2.

When a pulse is given via the "Drain" button, the vessel is emptied via valve V3 until LIS1 signals "EMPTY". (Note: The signal of the signal transmitters LIS1 and LIS2 is '1' when the transmitter is immersed in the liquid).

First consider how many memory elements (flip-flops) you need, by which signals the flip-flops are set, which signals reset the flip-flops and which actions must be executed when the flip-flops are set. Then complete the given logic diagram in Fig. 3.20 and plug the circuit onto the digital trainer.

Check whether the following conditions are met. Correct your circuit if necessary!

- Do not empty during the filling process, even if the corresponding button is pressed.
- When emptying, it is not possible to fill.

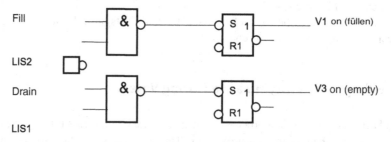

Fig. 3.20 Filling and emptying the tank. Circuit diagram for Exercise 3.15

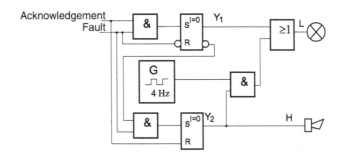

Fig. 3.21 Alarm with flashing signal for Exercise 3.17

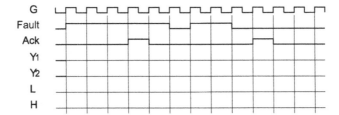

Fig. 3.22 Time diagram for Exercise 3.17

Exercise 3.16 (TANK32)

Now extend task 3.15 so that the vessel is automatically emptied as soon as it is full. To do this, you can use the signal LIS2 to set or reset flip-flops.

3.11   Alarm Circuit 3

Exercise 3.17 (ALARM33)

The circuit in Fig. 3.21 works like alarm circuit 2 (Fig. 3.17) of Sect. 3.9, but a flashing signal is also generated.

   Analyze the function of the alarm circuit from Fig. 3.21 with the aid of the time diagram according to Fig. 3.22. Complete the time diagram!

Dynamic Memory Elements and Counters

4

Abstract

To summarize what we have learned so far about flip-flops: an electronic memory element with two stable states is called a flip-flop. By suitable control, the flip-flop can be steered into the respective other state. In this way, binary (digital) information can be stored.

In the previous chapter we described the flip-flops where the output was changed immediately after changing the R or S inputs. These are the non clock-controlled flip-flops (bistable elements).

In this chapter you will learn how a memory chip must be constructed if it can only be controlled at a very specific point in time. Among other things, these devices form the basis for circuits that can *count* individual, short-term events ("pulses").

4.1 Clock Edge Controlled Flip-Flops

It is often desirable not to switch the input information at the S or R input directly through to the output of the flip-flop, but in dependence on a third signal, the clock. During the inhibit phase of the clock, the flip-flop does not react. Only as long as the "active" clock state persists does the flip-flop respond immediately. Several toggles can also occur during the active phase of the clock. An example of this is the state controls (Sect. 3.8), particularly the write inhibit.

However, this clock state control is not sufficient in all applications. Sometimes it does not make sense to allow several switchovers during the active clock phase. One would like to have exactly one fixed *time* at which the changeover takes place. In this case, it is possible to achieve a very precise coincidence of the times of the various changeovers

© The Author(s), under exclusive license to Springer-Verlag GmbH, DE, part of
Springer Nature 2022
H.-J. Adam, M. Adam, *PLC Programming in Instruction list according to IEC
61131-3*, https://doi.org/10.1007/978-3-662-65254-1_4

Fig. 4.1 Write lock combined
with read lock

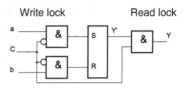

Fig. 4.2 Memory with
clock input

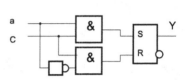

(synchronisation). Or unwanted disturbances can be suppressed in this way. The exact point in time can be the changeover of the clock signal, i.e. the transition of the clock from '1' to '0' or from '0' to '1'.

The set and reset inputs are then only used to prepare the storage. The store itself is only performed on a clock change when the clock goes either from '1' to '0' (falling clock edge, negative clock edge) or from '0' to '1' (rising clock edge, positive clock edge). We consider these clock edge-controlled flip-flops in the following section.

The edge-controlled flip-flops are built up from state-controlled flip-flops. We will study how this works in Example 4.1.

Example 4.1 (Write Lock Combined with Read Lock)

We combine the write inhibit with the read inhibit. To do this, consider the circuit in Fig. 4.1. When the clock is '1', the flip-flop can be read out, i.e. the flip-flop state Y' is present at output Y. During this time, the write inhibit prevents the flip-flop contents from changing, and the state is stable. If the clock is at '0' level, the flip-flop cannot be read; however, now the flip-flop can be set or reset. At the exact moment the clock transitions from '0' to '1' (i.e., on a rising edge), the read inhibit opens and the flip-flop contents are transferred to the output. We have thus achieved edge control.

But wait! Here we were a bit too fast: Check which state the output Y has during the '0'-phase of the clock! Correct, during this time there is always a '0' at the output, independent of the state Y' of the flip-flop!

The solution to this problem:

The signal Y' must not be sent directly to the output, but must be buffered. The circuit shown in Fig. 4.2, which we have already discussed in Sect. 3.8, is suitable for this purpose. ◀

Exercise 4.1 (RSFF41)

Connect the two subcircuits Figs. 4.1 and 4.2 to form a flip-flop circuit with dynamic input!

Fig. 4.3 RS bistable element
with dynamic input: Structure
from basic components

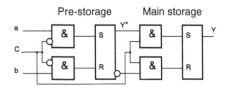

4.2   RS Bistable Element with Dynamic Input

If you have solved Exercise 4.1 correctly, you will recognize the circuit in Fig. 4.3.

Various flip operations can be understood by looking at Fig. 4.4: During the '0' phase of clock C, the write inhibit of the first flip-flop (pre-memory) is open and it is set or reset depending on the state of a and b; the input signals are initially stored in the pre-memory (Y^*). During the '1' phase of the clock, the write inhibit locks the pre-memory and the changes to the inputs have no effect. But now the write lock is open for the second flip-flop, the main memory, so that it takes over the state of the pre-memory (Y). This takeover happens exactly at the moment of the transition of the clock from '0' to '1' (rising edge).

In the '0' phase of the clock, the write inhibit locks the main memory: the output signal is retained. Because the pre-memory cannot change during the '1' phase of clock C and the main memory does not change during the '0' phase, changes to the output only take effect at the exact moment of the rising edge of the clock signal.

The information stored in the pre-memory is therefore transferred to the actual memory exactly when a rising edge of the clock signal occurs. Output Y therefore changes at a precisely predeterminable time! This process is called "triggering". (trigger on a rifle or shutter release on a camera). In the circuit symbol, the trigger property is represented by the small triangle at the input. In Fig. 4.4 you can clearly see that output Y^* "follows" the pre-memory output exactly at the time of the active rising edge, i.e. it is changed at exact points in time.

Fig. 4.4 Time diagram for the
toggle circuit Fig. 4.3

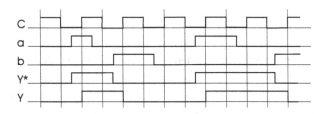

Fig. 4.5 RS bistable element triggered with the rising edge: circuit diagram and function table

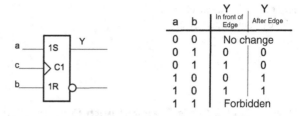

a	b	Y In front of Edge	Y After Edge
0	0	No change	
0	1	0	0
0	1	1	0
1	0	0	1
1	0	1	1
1	1	Forbidden	

Fig. 4.6 RS bistable element falling edge triggered: circuit diagram and function table

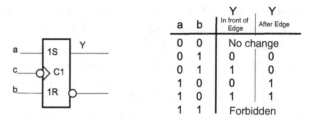

a	b	Y In front of Edge	Y After Edge
0	0	No change	
0	1	0	0
0	1	1	0
1	0	0	1
1	0	1	1
1	1	Forbidden	

Exercise 4.2 (RSFF42)

Build a dynamic RS flip-flop with the digital trainer using two RS flip-flops and test the function (Figs. 4.3 and 4.5).

If the clock signal is inverted, the flip-flop reacts to the transition from '1' to '0', i.e. to the falling edge of the clock signal (Fig. 4.6).

Exercise 4.3 (RSFF43)

Using basic building blocks (AND elements and RS flip-flops), draw the logic diagram of a dynamic RS flip-flop with falling edge control.

4.3   The JK-Flip-Flop

The following applies to the circuit just discussed: At the time of the active edge, it must be ensured that the set and reset inputs (*a* and *b*) do not carry a '1' signal at the same time. This restriction can be lifted by additional circuits, which we do not want to go through here.

The JK-flip-flop is a dynamic RS bistable element which does not assume an undefined state due to an additional circuit (Fig. 4.7). The combination '1', '1' at the preparation inputs is therefore no longer prohibited. The state J and K both equal to '1', lets here the output toggle at each active edge. The toggle shown here switches on the rising (positive) edge of the clock signal. The J input corresponds to the set input, the K input to the reset input. The letters J and K have no relation to the function of the flip-flop, they were chosen arbitrarily.

Fig. 4.7 JK bistable element edge-triggered: Circuit diagram and function table

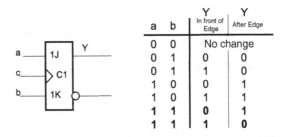

a	b	Y In front of Edge	Y After Edge
0	0	No change	
0	1	0	0
0	1	1	0
1	0	0	1
1	0	1	1
1	1	0	1
1	1	1	0

Exercise 4.4 (JKFF41)

A negation sign before the clock input of the symbol means switching with the falling (negative) edge.

Draw the symbol of a JK bistable element with falling edge control and create the function table. Check the function of a JK bistable element with the digital trainer.

4.4 The T-Flip-Flop

The JK flip-flop can be wired to flip on each active edge. If you ask yourself now, what such a circuit is good for, we can only put you off until later, when we discuss counter circuits further down. Because of the application to counters this property is even so important, that there is a special circuit for it, which works exactly like the JK-flip-flop with J and K equal to '1', but does not have the two inputs at all, but has only a single clock-input.

With the **T-Flip-Flop**, the output variable changes with *each* active edge of clock pulse C. ("T" from the english word "*toggle*").

Exercise 4.5 (TFF41)

Complete the function table in Fig. 4.8! Draw the symbol for a falling edge-controlled T- Flip-Flop. Connect a JK Flip-Flop to the digital trainer as a T bistable element and test the function.

Fig. 4.8 T Flip-Flop for Exercise 4.5

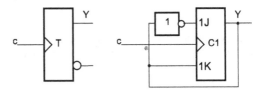

The T-flip-flop is easier to handle for some special use cases because it has no unneces-sary functions, and it is cheaper because it needs less connection feet. But as it is with specialists: they are often too one-sided. The JK flip-flop is more versatile, and is often given additional R and S inputs to allow the flip-flop to be set and reset independently of the clock.[1]

4.5   Automatic Filling and Emptying of a Measuring Vessel

As an example, let us fill and empty the measuring vessel in Fig. 4.9. In the circuit shown in Fig. 4.10, a T-flip-flop is set with the start button. Valve V1 for filling the measuring vessel in Fig. 4.9 is opened and remains open until the upper limit switch LIS2 gives a signal. Then valve V1 closes and valve V3 opens. The measuring vessel runs empty until the signal from the lower limit switch LIS1 is logically 0. The sequence stops until a press on the start button restarts the cycle. With each press of the button, the quantity of the measuring vessel is therefore dispensed exactly once.

Exercise 4.6 (TANK41)

Build the circuit from Fig. 4.10 and test its function! What change in the circuit and in the operation results if an "ordinary" RS flip-flop is used instead of the T-flip-flop? What is the advantage of using the T-flip-flop in conjunction with the start button?

Fig. 4.9 Measuring vessel

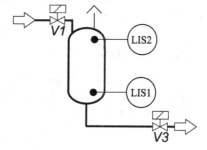

---

[1] Kits for training usually have these types built in. They are usually designed so that you can just leave the inputs you don't need open (i.e., unconnected).

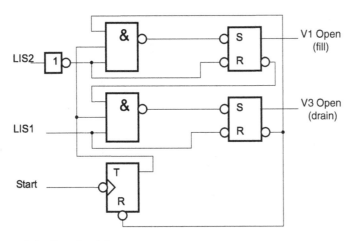

Fig. 4.10 Circuit for Exercise 4.6

4.6 Counter

The T-flip-flop discussed in the previous section is suitable for setting up digital counters. Let us now examine this in a little more detail: In digital counters, each bit has a valence in a power of two, according to the definition of dual numbers. We have already seen this at the very beginning of this course in Sect. 1.6.

Table 4.1 shows the numbers 0–15 in decimal form, 0–F in hexadecimal form and 0000–1111 in dual form. The dual numbers are 4 bits wide. A digital counter must have one output for each bit of the number to be displayed. In addition, a clock input is required whose pulses increment the counter one value at a time. The outputs must pass through the states in sequence, as shown in the table. Each clock pulse toggles the outputs (i.e. the positions for 2^3, 2^2, 2^1 and 2^0) as required for the number in the line below.

Plotting this as a line graph against time gives Fig. 4.11. Focus first on the top two lines showing the signal waveforms for the *clock* and the 2^0 signal. At the end of each clock there is a "falling edge" of the clock signal; i.e. the clock goes from '1' to '0'.

At this exact time, the signal for 2^0 must change state. So note that on each *falling* edge of the clock, the 2^0 signal toggles.

Now also look at the line for the 2^1 signal and compare it with the 2^0 signal. The 2^1 signal flips on each falling edge of the 2^0 signal. This behaviour corresponds exactly to the transfer of a full bundle of two to the next higher digit described in Sect. 1.6.

Dec	HEX	Dual			
Table 4.1 Decimal, HEX and dual numbers		2^3	2^2	2^1	2^0
0	0	0	0	0	0
1	1	0	0	0	1
2	2	0	0	1	0
3	3	0	0	1	1
4	4	0	1	0	0
5	5	0	1	0	1
6	6	0	1	1	0
7	7	0	1	1	1
8	8	1	0	0	0
9	9	1	0	0	1
10	A	1	0	1	0
11	B	1	0	1	1
12	C	1	1	0	0
13	D	1	1	0	1
14	E	1	1	1	0
15	F	1	1	1	1

So it goes on: at each falling edge the "following" signal toggles. So to build a counter, flip-flops are needed, which flip every time the clock has an active (here falling) edge. Aha, we see that here the JK-flip-flop or the T-flip-flop will be applied.

4.6.1 The Asynchronous Counter

Switching elements which flip at each active edge and can therefore be used in counting circuits are the T flip-flop and the JK flip-flop with $J = K = 1$. These flip-flops change the output level at each active edge of the input clock.

For the described counter application the falling edge must be the active clock edge. If you connect four T-toggle elements in series as shown in Fig. 4.12, you will obtain a 4-bit binary counter.

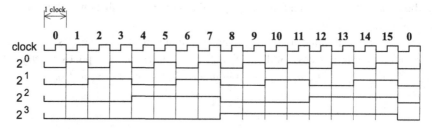

Fig. 4.11 Timing diagram for 4-bit counter

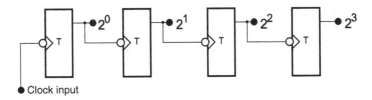

Fig. 4.12 Asynchronous 4-bit counter with T-toggle elements

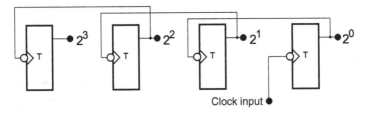

Fig. 4.13 Asynchronous 4-bit counter with T-toggle elements, alternative representation

Exercise 4.7 (COUNT41)

Describe the operation of the counter circuit Fig. 4.12. In particular, describe the switching of the flip-flops when all four outputs indicate '1' and another clock pulse arrives.

The falling clock edge causes the first flip-flop to toggle, it is set. A rising edge is therefore produced at the output. This rising edge of the output 2^0 has no effect on the following flip-flop. Only the next falling(!) clock edge causes the first flip-flop to flip back. This reset affects the second flip-flop (2^1) as a falling edge, so it now flips afterwards as well. You can see that the individual flip-flops switch one after the other and not all at the same time, i.e. they do not work synchronously. This counter is therefore called an "asynchronous counter".

In the drawing Fig. 4.12 it is unattractive, that the least significant output is drawn at the left and the most significant output at the right. In the positional notation of the dual numbers, however, it is just the other way round: the most significant digit is in the first position, i.e. on the left, and the respective lower ones follow after the right. If we represent the asynchronous counter as in Fig. 4.13, the order is correct again.

Exercise 4.8 (COUNT42)

Build a dual counter in expansion level 4 (i.e. 4-level, 4-bit) with the digital trainer. Use JK flip-flops that you switch as a T-flip-flop.

Fig. 4.14 Asynchronous 4-bit down counter

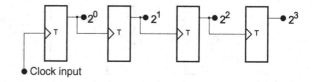

Fig. 4.15 Asynchronous 4-bit down counter, alternative circuitry

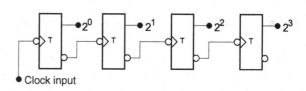

Fig. 4.16 Electronic dice

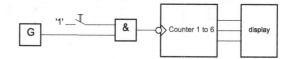

4.6.2 The Asynchronous Down Counter

A down counter is obtained by either using *rising-edge triggered* toggle elements (Fig. 4.14), or by triggering the next higher position with the respective *negated* outputs in the case of falling-edge triggered toggle elements (Fig. 4.15).

Most of the toggle elements available in practice have additional R and S inputs, which can be used to set or reset the flip-flop independently of the clock signal.

Exercise 4.9 (COUNT43)

Draw the line diagram for the signal waveforms for a down counter. Use the table of numbers in Sect. 4.6 to help you. Use the digital trainer to build a down counter and test it.

4.6.3 Modulo-n, Decimal- and BCD-Counters

Perhaps you can already think of some applications where you can apply your newly acquired knowledge of counters. However, before we look at applications, we would like to present you with some more theory.

Imagine you want to build an electronic dice. It is to output the numbers 1–6. In the circuit Fig. 4.16, G is a generator which supplies a high-frequency ("fast") clock signal. This reaches the counter via the gate circuit (AND element) as long as the key is pressed. The clock is sufficiently fast that the number of cycles until the key is released is unpredictable; the counter shows a "random" number.

Fig. 4.17 Circuit for
Exercise 4.10

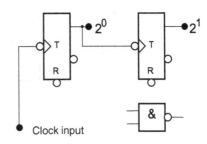

Clock input

4.6.4   Modulo-3 Counter

But before we build a 1–6 counter, we start with something simpler, a counter modulo 3
with the counting sequence 0,1,2,0,1,2..., It must display the binary numbers 00, 01 and
10 one after the other. The fourth number 11, which can be displayed with two bits, must
not appear, but must be immediately reset to 00 in its place. This can be achieved by immediately resetting the flip-flops via the R inputs when this "wrong" number appears. Thus,
in practice, the undesired number appears only for such a short time that it is not noticed.
The reset signal must appear at the bit sequence $11_{dual} = 3_{dez}$. Since the flip-flops are usually
reset with a '0' at the R input, the reset signal can be generated with a NAND element.

Exercise 4.10 (COUNT44)

Complete the circuit diagram in Fig. 4.17 to a counter modulo 3, i.e. it has the counting
sequence 0, 1, 2, 0 ... and build it with the digital trainer!

Exercise 4.11 (COUNT45)

Build a counter modulo 6 (0 ... 5) with the digital trainer!

4.6.5   BCD Counter

In Sect. 1.12 we have already introduced you to BCD numbers. Digital counter circuits
that display decimal numbers in BCD format "directly" need a modulo-10 counter for
each of the decimal places. Three flip-flops are not enough for this. With four flip-flops,
that is four bits, a counter can count from 0 to 15 (= 16 values). But to display the BCD
numbers, you only need the numbers 0–9. A counter, which should only display the decimal places 0–9, must switch back to "0" after displaying the digit "9", not to the next
digit "A".

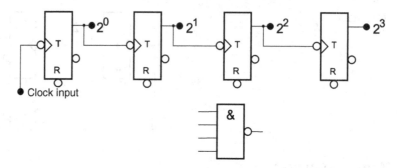

Fig. 4.18 Circuit for Exercise 4.12

Fig. 4.19 Circuit for 1 … 6 counter for Exercise 4.13

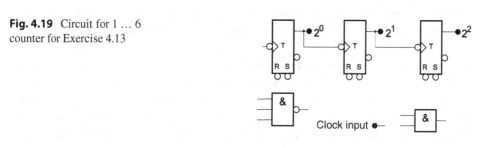

Exercise 4.12 (COUNT46)

Complete the circuit diagram according to Fig. 4.18 to a counter modulo 10 (0 … 9) and build it up with the digital trainer! Justify why instead of the NAND element with 4 inputs one with only 2 inputs is sufficient!

4.6.6   Counter with Any Start and End Value

The counter circuits considered so far always started from the number 0. But for the dice mentioned at the beginning, the counter must start at the number $1_{dez} = 001_{dual}$ and count up to the number $6_{dez} = 110_{dual}$. You can realize this by not simply resetting all flip-flops to 0 when the "wrong" number $7_{dez} = 111_{dual}$ is reached, but by using the R and S inputs to reset/set the flip-flops so that the desired number is displayed $1_{dez} = 001_{dual}$ after the clock edge.

Exercise 4.13 (COUNT47)

Modify the circuit of the previous Exercise 4.12 so that the counting sequence 1 … 6 is created! Add a gate circuit to the clock input in Fig. 4.19 so that you obtain an electronic "dice"!

Fig. 4.20 Process model for
Exercises 4.14 and 4.16

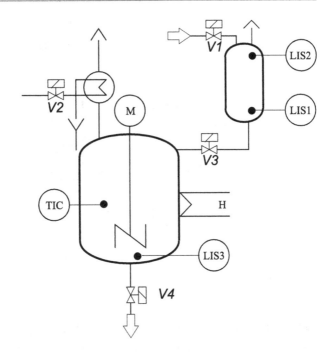

4.7 Multiple Filling and Emptying of a Vessel

If a vessel is to be filled several times in succession, each filling of the vessel can generate
a pulse which is registered by the counter. The counter reading reached indicates the number of fillings.

Exercise 4.14 (TANK42)

Describe the function and operation of the circuit in Fig. 4.21 that controls the process
in Fig. 4.20. Build the circuit and test its function.
Expand the circuit so that the filling process is not only two, but three times in succession (Fig. 4.21).

4.8 Timer

Each counter circuit requires a certain time until the entire counting cycle has been run
through once, i.e. the counter end value has been reached. You can calculate this time by
multiplying the counter reading by the duration of a clock pulse. With a cycle of $f = 1$ Hz,
i.e. one pulse per second, it takes 180 s (= 3 min) until the counter has reached the level
10110100_{dual}, correspondingly 180_{dez}. So you need 8 flip-flops for a 3 min counter with a

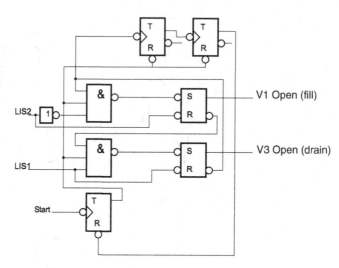

Fig. 4.21 Circuit for Exercise 4.14

clock duration of 1 s. If you reduce the clock duration, you need more flip-flops, but the accuracy with which you can set times increases by the same amount.

Exercise 4.15

How many flip-flops do you need to build a counter for a count time of 3 min if the clock pulses have a duration of 0.5 s? How many are required for a clock duration of 0.1 s?

Exercise 4.16 (MIXER41)

After pressing a start button, the stirrer M should run for 4 s (Fig. 4.20). Draw the corresponding logic diagram and build the circuit with the digital trainer.

Part II

PLC Technology

In the second part of the book we cover PLC technology. You will now quickly see the advantage of PLC programming: Changes don't require cumbersome cable pulling, as was necessary with digital circuitry. You simply write the new instruction and that's it.

Programming should be learned in the most universal language, the *instruction list*. Very often, graphical languages (e.g. function block diagrams) are preferred to textual languages because of their visually appealing presentation. The program representation then resembles the (familiar) circuit diagrams or circuits with digital switching elements. However, one has to be careful here: the seemingly so easy programming by means of graphical representation is no longer so trivial in the case of switching devices, i.e. when *memory and time elements* occur.

Combinatorial Circuits with PLC

5

Abstract

In the case of logic switching elements, it was clear without any special mention that the output (more precisely, the output signal) depends on the inputs (the input signals). The relationship, i.e. the function of the circuit, is clearly defined. You get control when you connect several switching elements together. The signals at the inputs cause the signals that can be tapped at the outputs. The function of the overall circuit is determined by which switching elements are connected to each other and how. The function can be changed at any time by inserting or removing switching elements or changing the connections. In a PLC, as in digital circuits, we also have inputs and outputs. The dependence of the output signals on the input signals is not determined by the wiring, but it is determined by a "program" which signals are to be linked together. This program can be written like an ordinary text with the PC. The advantage of this technique is the standardization of the controls. Because only a few universal types of control devices are required, they can be produced in large quantities. Only the buyer decides by programming which of the "inner building blocks" are to be used and how they are to be "wired".

For the practical application, one can initially assume that an "infinite number" of individual modules are available. The "wiring" is done by the program. Another advantage is that the function can be changed at any time by changing the program. A screwdriver or soldering iron is not required for this.

We will first practice programming using the "Instruction List". In this *instruction list (IL)*, the individual links are described by short texts. Each line of the IL contains one instruction. Each instruction consists of an operator and an operand. The operator

© The Author(s), under exclusive license to Springer-Verlag GmbH, DE, part of
Springer Nature 2022
H.-J. Adam, M. Adam, *PLC Programming in Instruction list according to IEC
61131-3*, https://doi.org/10.1007/978-3-662-65254-1_5

Fig. 5.1 Binary inputs and
outputs: direct addressing

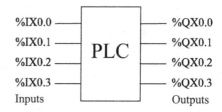

determines the type of link (e.g. OR link), the operand determines the input/output
terminal whose signal is to be linked.

5.1   Directly Represented Variable

In the program you must be able to address the inputs and outputs of the PLC. The IEC
standard prescribes very specific identifiers when the addresses of the hardware are to be
addressed. Some examples can be found in Fig. 5.1. The identifier for the direct addresses
starts with the percent sign '%', followed by the letters 'IX' for an *input* bit or by the letters
'QX' (*to quit*) for an output bit. This is followed by the number of the terminal.

Examples

%IX0.1 or %QX0.0.

Other designations are used for other input or output types. But of it later. At first we only
care about single logical input/output bits. By the way, you can leave out the 'X' for these
types for simplification. It is also irrelevant whether the identifiers are written with upper
or lower case letters.

Valid Identifiers Are Therefore

%IX1.4, %I4.0, %Q1.1, %qx2.0, %i0.1, %iX1.2, %q3.4, etc.

> Directly represented *variables* begin with the % character, directly represented *inputs*
> begin with %IX, and directly represented *outputs* begin with %QX.

5.2   Basic Logical Links with PLCs

Of course, a PLC is "underpowered" if it only has to implement a single link. But we have
to start with simple exercises, therefore we will first create only the already known
basic links.

If you are looking for an *overview of* the links possible with the PLC, you will find it in this book in Chap. 16.

5.3   OR Operation

Example 5.1 (Lamp Circuit with OR Operation)

A lamp should be switched on when the infrared detector has responded or a button is pressed.

Clearly, the OR operation (Fig. 5.2) must be used here! In the digital technology part (Sect. 2.8) you have already learned how to translate this task into a function table. ◄

To have the PLC perform this function, you must do two steps:

- first, connect the infrared detector and the pushbutton to the *input terminals* of the PLC and connect the lamp to an *output terminal* of the PLC; and
- secondly, write the *instruction list* so that the PLC works as an OR element.

So on the one hand you have to connect the process to the PLC control device and on the other hand you have to write the control program.

The symbolic variables *a*, *b* and *x* are assigned to the real signals, and these in turn are connected to the input/output terminals of the PLC. This can be clearly displayed in an "assignment list".

```
Assignment of the inputs/outputs:
Inputs:
Variable a (infrared detector) %IX0.0
Variable b (pushbutton) %IX0.1
Exit:
Variable x (lamp) %QX0.0
```

These assignments are written in the instruction list before the actual program lines. The assignment list ("terminal diagram") is introduced in the program header with the keyword 'VAR' and concluded with the keyword 'END_VAR'. You can choose any names for the variables in the assignment list (as long as they contain only letters, digits and the underscore (_) and do *not* begin with a digit). After the keyword 'AT' you enter the direct

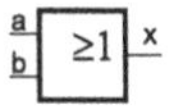

Fig. 5.2 OR link

$$x = a \lor b$$

a	b	x
0	0	0
1	0	1
0	1	1
1	1	1

PLC address. After this you put a colon followed by the variable type. Initially, we only use individual bits; the type designation for this is 'BOOL'.

The symbolic identifiers (variable names), their type and terminal address are defined in the program header. For PLC practice, this means that you have a kind of "wiring diagram" (assignment list) integrated in the program.

```
VAR
  a AT %IX0.0: BOOL
  b AT %IX0.1: BOOL
  x AT %QX0.0: BOOL
END_VAR
```

After the header the actual program lines are written. At the beginning of a link, the value of the first variable is loaded. The operator for this is called LD (*to load*). In the example, the next value is to be linked with the just loaded "OR". The operator OR is provided for this purpose. The result of this link can then be assigned to the output with the ST operator (to store).

```
LD a
OR b
ST x
```

The LD operator loads the first operand from the input. At the same time a new instruction sequence is started.

The OR operation is created by the OR operator. The value loaded from input '*a*' is linked with the value present at input '*b*'.

The assignment of the logic result to the output is done with the ST operator.

5.4   The Current Result

Have a little patience before you create the first program! You should be very clear about how the PLC works: the LD operator loads the first operand '*a*' into a very specific memory location, the so-called *"current result"* (CR).

The Current Result CR is a memory location that contains *one* operand.

The next operator OR links the value of the "current result" with its operand *b* and overwrites the memory location "current result" with the result of this operation. Aha, is the name of this memory location clear now?

After each arithmetic operation the memory location CR is overwritten with the result. The last instruction in this program, the ST instruction, does not change the "current result", but copies the value to the memory location *x* specified in the operand. Note that after the ST instruction is executed, the current result retains its value and can be used further.

In Summary

- With 'LD' the current result CR is set completely new, independent of the previous value. A new sequence starts.
- The 'OR' operator changes the current result; the new value of the AE depends on both the previous value of the CR and the operand specified after the operator.
- 'ST' does not change the CR, it can be reused in subsequent operations.

However, the program is not yet finished. In the instruction list according to IEC 61131-3 some formalities still have to be observed.

The program starts with the keyword PROGRAM <ProgramName> and ends with END_PROGRAM.

There is only one instruction in each line of the instruction list. It begins with the operator for the function to be executed and an operand that indicates what the operator refers to. In most cases, the operand is the identifier for an input or output. By the way, the case does not matter in the whole text of the statement list and can be mixed arbitrarily.

Example 5.2 (Complete Program Code: OR51)

```
PROGRAM OR51
  VAR
  a AT %IX0.0: BOOL
  b AT %IX0.1: BOOL
  x AT %QX0.0: BOOL
END_VAR
  LD a
  OR b
  ST x
END_PROGRAM                                      ◄
```

5.5   The PLC Simulation Application "PLC-Lite"

There is one last hurdle to overcome before you can start programming. On the website[1] of the authors to this book, the application PLC-lite is available, with the help of which you can do all the exercises.

Installing and Starting PLC-Lite

For more information on installation and use, please refer to the app documentation. In this section, we give you a brief introduction to its use in connection with the exercises in this book.

   After starting the app, the application window opens with the controls for controlling the simulated PLC and the editor window where you can type in the instruction list. You open the models for processes and input/output elements with the "Visualization" menu.

Operation of PLC-Lite

In the main window (Fig. 5.3), click on the "Visualization" button (or "Processes" in older versions of PLC-lite) and first select "Standard I/O". You can use this menu to open the associated display elements and process models in later exercises.

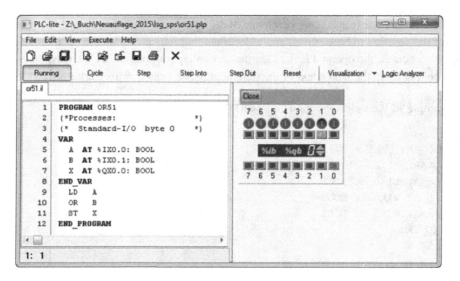

Fig. 5.3 PLC-lite

To set the input values for the PLC, click on the green buttons with the mouse. The lower, red "light emitting diodes" (LEDs) show you the output value. A "lit" LED means a logical '1'. The inputs act as buttons when you click the mouse. If you drag the mouse away from the button while holding down the left mouse button, it remains 'latched'. The input then remains permanently switched on until you click on the input again. Thus you have a "switch" at the input.

Exercise 5.1 (OR51)[2]

Realize an OR operation with the PLC. Create the instruction list with the PLC-lite app and test the program. The test is positive if the function table for the OR operation is fulfilled for all switch positions.

The simulation is started with "Run". The button then changes its status to "Running"; clicking on it again stops the simulation. "Reset" completely resets the simulation including all process models.

After clicking the Run button, the system checks the statement list. If errors are detected, a corresponding message is displayed. Correct the errors. The simulation will not start until there are no more errors in the statement list.

Further Possibilities of Simulation

A useful function is the *single step mode*. A click on 'Step' executes the next instruction in the instruction list and then waits. A click on 'Cycle' works through all instructions once and stops at the beginning of the instruction list.

The instruction that will be executed in the next step is highlighted in the editor. In the simultaneously opening window 'Watch expressions' the current values of the variables and the current result (CR) are displayed. Here are two columns for the values: First look at the column 'Value' only. We will explain the meaning of 'Direct Value' in Sect. 6.4.

Finally, save the file first, then save the project as well. This 'two-step' approach is important for later, when a project will have more than one instruction list. Note therefore: 'Save file' saves the current statement list. The project as a whole must be saved separately with 'Save Project'.

---

[2] For many exercises, you can find suggested solutions on the authors' website under the name given in parentheses after the exercise number.

If you get "stuck" somewhere, you can also find the solutions to the exercises at
www.adamis.de/plc/.

5.6   AND Operation

Whew!! That was a nice bit of work for such a simple function! But don't worry, you can
take advantage of the programming right away when you tackle the next example:

Example 5.3 (Darkroom Lighting)

In a photo lab room, the yellow light may only be turned on when the door is locked
and the windows are darkened. ◄

In Part I (Fig. 2.3) you have already learned how to translate this task into a function
table. You get the AND operation (Fig. 5.4).

Again, you must first connect the input terminals of the PLC to the door or darkening
contacts and the lamp to an output terminal of the PLC, and secondly write the instruction
list to make the PLC work as an AND gate.

In our example, a door contact is to emit the logical signal '1' when the door is closed.
When the window darkening is effective, the window contact emits signal '1'. The lamp
with the yellow light is switched on when it receives signal '1'. The symbolic variables a,
b and x are again assigned to the real signals. This results in the assignment list:

```
Assignment of the inputs/outputs:
Inputs:
Variable a (door contact) %IX0.0
Variable b (window contact) %IX0.1
Exit:
Variable x (yellow light) %QX0.0
```

Just as in Example 5.1 with the OR operation, the value of the first variable is loaded
(LD). The second variable must be "AND"-linked with the just loaded one. The operator
'AND' is provided for this purpose. The resulting current result is again assigned to the
output with the ST operator.

Fig. 5.4   AND operation

$$x = a \wedge b$$

a	b	x
0	0	0
1	0	0
0	1	0
1	1	1

The AND operator is used for the AND operation.

Can you see that this results in practically the same program as in the previous Exercise 5.1? You only need to change the OR operator into the AND operator! No screwing or soldering of cables, no changing of components …

Exercise 5.2 (AND51)

Realize the circuit for the photo lab by an AND operation! Create the instruction list, then program the PLC and test the program. The test is positive if the function table of the AND operation is fulfilled for all switch positions.

5.7   Negation of Inputs and Outputs

The logical values of the inputs and outputs can be negated before further processing (Fig. 5.5). This allows negators as well as NAND and NOR links to be implemented. The negation of the operand is generated by the modifier "N" (not). The modifier is appended to the operator. Thus we get the modified operators: LDN, ANDN, ORN, STN.

Example 5.4 (Showroom Monitoring)

With the help of a capacitive encoder, a sales room monitoring is to be carried out. The encoder gives a logical '0' if there is a customer in the room. In this case, an optical signal is to light up. ◄

```
Assignment of the inputs/outputs:
Entrance:
Variable a (capacitive encoder) %IX0.0
Exit:
Variable x (signal lamp) %QX0.0
```

As a solution for the example simply results in a negation. Depending on whether you negate the input or the output, you get different programs: in the first case the LD operator is modified, in the second the ST operator.

Fig. 5.5 Negation of input resp. output

Example 5.5

PROGRAM NOT1	PROGRAM NOT2
VAR	VAR
a AT %IX0.0: BOOL	a AT %IX0.0: BOOL
x AT %QX0.0: BOOL	x AT %QX0.0: BOOL
END_VAR	END_VAR
LDN a	LD a
ST x	STN x
END_PROGRAM	END_PROGRAM

◀

In this case it does not matter whether the LD operator (input) is modified with negation or the ST operator (output).

In addition to the LD and ST operators, the AND and OR operators may also be modified with N. Thus NAND and NOR links can be realized. Exercise 5.3 can be solved with a NAND link. The negation is at the output; you must modify the ST operator to the STN operator.

Exercise 5.3 (MIXER51)

A valve must not be open ($x = 0$) when the stirrer ($a = 1$) and the heater ($b = 1$) are switched on. Write the program as an instruction list and draw the function chart.

5.8 Boolean Algebra: De Morgan's Rules

Let us now examine what difference results when the LD, AND or OR operator (i.e. the input) or the ST operator (i.e. the output) is modified with negation. You should now have a look at Sect. 2.20! The relation given there is called "de Morgan's rule". It specifies how to "bring forward" the negation from the output to the inputs. Of course the operation changes: AND becomes OR and vice versa! Compare the logical switching functions with the PLC program codes and the switching symbols!

Exercise 5.4 (NAND51)

Implement and test a NAND logic in the two variants from Table 5.1 on the left.

Exercise 5.5 (NOR51)

Test also the (real) NOR-logic and its realization as AND-logic with negated inputs!

Table 5.1 de Morgan's rules

NAND		NOR	
$\overline{a \wedge b}$	$\overline{a} \vee \overline{b}$	$\overline{a \vee b}$	$\overline{a} \wedge \overline{b}$
LD a	LDN a	LD a	LDN a
AND b	ORN b	OR b	ANDN b
STN x	ST x	STN x	ST x

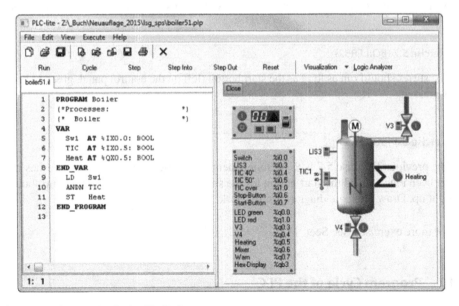

5.9   Boiler Heating (Two-Point Control)

A reaction vessel is to be kept at a constant temperature via a contact thermometer as a limit signal transmitter as long as the heating is switched on. The temperature controller TIC emits the signal '1' as soon as the set temperature (e.g. 50 °C) is reached or exceeded. It should be possible to switch the heating on and off with a switch. This "main switch" emits signal '1' when it is closed (Fig. 5.6).

Fig. 5.6 Boiler heating in the "Boiler" process

Fig. 5.7 Circuit for
Exercise 5.6

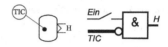

Exercise 5.6 (BOILER51)

Implement and test the heating control with PLC according to the circuit Fig. 5.7.

```
Assignment of the inputs/outputs:
Inputs:
Switch Sw1 %IX0.0
Thermometer TIC %IX0.5
Exit:
Heating Heat %QX0.5
```

Use the "Boiler" process to simulate (Fig. 5.6). The switch for switching on and off is
located on the far right of the field with the controls.

In the process, you will find "manual buttons" next to the valves V3 and V4 as well
as next to the heater, with which you can switch these elements *directly,* without being
influenced by the program. By actuating the heater, you can also cause a fault due to
overheating, for example.

Do not be disturbed by the rapid switching on and off of the heating when the temperature is exceeded. We will later learn ways to "calm down" the control. Also try out
what happens when you overheat the boiler by clicking on the "Heating" manual button.

Exercise 5.7 (BOILER52)

The stirrer should run as long as the start button (left on the control panel, green button)
is pressed.

Exercise 5.8 (BOILER53)

The previous Exercise 5.7 is to be supplemented as follows: If the level LIS3 has not
yet been reached and the start button for the stirrer is pressed, the warning lamp should
light up. Draw the logic diagram for this task!

For more exercises, see Sect. 13.2.

5.10   Program Cycle of the PLC

Now that you are a little familiar with PLC programming, let's take a look at how the PLC
controller works. If you think back to Exercise 5.8, you can easily see that the instruction
to query the start button must be executed not just once, but repeatedly, because the program is supposed to respond to the button press at all times.

Fig. 5.8 Cyclic processing of
the control program
with a PLC

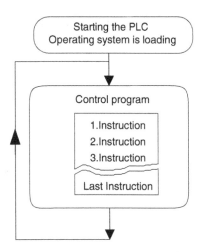

In fact, this is also the case with PLC controllers: the instructions from the instruction list are not processed only once, but repeatedly "in a circle". After the last instruction has been processed, the process starts again from the first one. Each run is called a "cycle". This is graphically represented in Fig. 5.8 by the arrows in the connecting lines.

> The PLC program starts with the first command at startup and processes all command lines one after the other until the last command. After that, the cycle always starts again with the first command.

Often the next cycle is not started immediately after the end but the cycles are started by a timer in certain time intervals. You can set this time in PLC-lite.

This cyclic mode of operation ensures that the inputs are regularly polled again and again and that any changes made in the meantime are registered.

5.11 Circuit Diagrams

Electricians understand the circuit diagram immediately. You may be a "non-electrician", but you can certainly understand how the electrical circuit in Fig. 5.9 works: current can flow downwards from the upper "busbar" via wires or closed switches, as in a watercourse. Open switches interrupt the current flow. That's not so difficult, is it?

Fig. 5.9 Circuit diagram and function chart

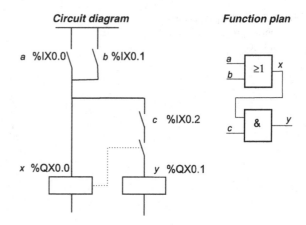

With this approach, it is no longer a problem to recognize that the current can always flow through two switches connected in parallel when at least one of the two switches is closed. So it is an OR-circuit. This is the case with the two contacts that are to be connected directly to the PLC inputs %IX0.0 or %IX0.1.

Current can only flow via the switch to be connected to %IX0.2 if output %QX0.0 is active. This is therefore an AND circuit: output %QX0.1 is '1' if output %QX0.0 and input %IX0.2 are both '1'.

This electrical circuit is shown in Fig. 5.9 as a functional diagram.

5.12   Querying Output Variables

Frequently, the control of other variables depends on the status at one or more outputs. The PLC must therefore be able to query an output and process its logical state. This is possible without any problems.

But Note: the output is still an output, you cannot use it in the plant "like an input" and feed signals into the PLC via it, but you can only read out the last value stored to this output in the program.   In the heating control example, you could still switch on a pilot lamp while the heating is running. In program Example 5.6 you can see the program extension. Note how the *output* "Heat" is queried with the LD command.

Fig. 5.10 Circuit for
Exercise 5.11

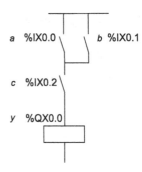

Example 5.6 (Boiler: Control Lamp)

```
PROGRAM NOT3
VAR
 . . .
 Heat AT %QX0.5: BOOL
 On_LED AT %QX0.0: BOOL
END_VAR
 . . .
 LD
 ST On_LED
 . . .
END_PROGRAM                                                    ◀
```

Exercise 5.9 (BOILER54)

Complete the program from Exercise 5.8 with the control lamp as described in
Example 5.6!

Exercise 5.10 (POWER51)

For the circuit in Fig. 5.9, assign the variables correctly, create the instruction list and
implement the circuit with PLC-lite.

Exercise 5.11 (POWER52)

Program the circuit according to Fig. 5.10. Compare the result with the following
Example 5.7 and Eq. (5.3)!

5.13 Flags and Brackets

Example 5.7

Exercise 5.11 can be solved by the following program part:

```
LD  %IX0.0
OR  %IX0.1
AND %IX0.2
ST  %QX0.0
```
◀

The OR statement for inputs %IX0.0 and %IX0.1 is executed first, and the result is written to the CR, which is then linked to %IX0.2. As the last final result, the CR is written to output %QX0.0.

As a mathematical function equation, this relationship can be stated like this, where we introduce x as an intermediate result of the OR operation in order to be able to represent the two steps separately:

$$x = a \vee b \tag{5.1}$$
$$y = x \wedge c \tag{5.2}$$

or, x from the first equation inserted:

$$y = (a \vee b) \wedge c \tag{5.3}$$

If you don't use x as an intermediate result, but put the corresponding term directly into the equation for y, you can "save" an output. And yet a memory location is required:

The parenthesis in the formula means that the expression for a and b in the parenthesis is evaluated first, and this result (stored in the CR in this particular case) is associated with the value c.[3]

5.14 Memory for Flags

Sometimes one does not want to lose the intermediate result x of Eq. (5.2). Then it is not enough to use CR for storage. We have to use additional memory space. First, we apply a common procedure in the PLC to get us memory space.

---

[3] In the instruction list of the PLC, such a bracket expression can be programmed directly. But that comes a little later in Exercise 5.12.

One could assign the intermediate result *x* to an output. But it can also be done without wasting the usually rare output terminals. Consider Example 5.8: We use a special memory location for intermediate storage, a *memory-flag* x AT %MX0.0: BOOL, which can hold the intermediate result *x*. By using a memory location (flag), the value is temporarily stored, "remembered".

The result of the expression for *x* in the parenthesis is (a $\vee$ b) assigned to a "*flag*" x with the ST operator and can be retrieved at any time by querying this flag. You can give flags symbolic names in the variable list.

Example 5.8

```
PROGRAM Memory
VAR
  a AT %IX0.0: BOOL
  b AT %IX0.1: BOOL
  c AT %IX0.2: BOOL
  x AT %MX0.0: BOOL (* flag *)
  y AT %QX0.0: BOOL
END_VAR
  LD a
  OR b
  ST x (* save to flag *)
  LD x (* read from flag *)
  AND c
  ST y
END_PROGRAM                                                      ◄
```

5.15   Comments in the Instruction List

In the instruction list (Example 5.8) you will find texts framed by the strings "(*" and "*)". These texts are comments and serve to document the program. You can significantly increase the readability and comprehensibility of the program by using clever comments. These comments are intended as information only for yourself, so that you can still understand the program run (even after some time!). The comments have no influence on the actual program run of the PLC!

Flags are given symbolic names. The memory location is addressed directly with %MXa.b.

- Flags are nothing more than memory locations for a value that you can recall at any time.
- Flags have in common with outputs that you can assign a value to them and read it out again at any time. Unlike these, however, they have no connection to the outside.
- Because absolute memory space is practically never important for flags, you can confidently leave the allocation of memory space to the operating system. In the variable list, specify only the symbolic name and the type "BOOL" without assigning a special memory location. Instead of x AT %MX0.0: BOOL, just write x: BOOL. The actual memory location is automatically determined by the PLC system.

5.16   Intermediate Results in Brackets

The programming of the brackets is initially very pleasant for those who are mathematically experienced, because it looks similar to the familiar rules from mathematics. An example can be found in the following Exercise 5.12 (program 3).

By modifying the operators with the ' ('modifier and the') ' operator, the expression can be written as Instruction List.

The modifier (after the AND and OR operators creates the AND (and OR (operators, which begin a parenthesized expression.

The operator) "Close parenthesis" closes the expression and causes it to be processed.

In PLC practice, however, the representation with the brackets quickly becomes confusing. In many cases, the intermediate result is already available at an output and can be used directly from there by querying the output. In general, intermediate results will be assigned to flags. At first, programming with flags seems more effortful than using brackets: the program gets more lines, and the way of thinking is unfamiliar if you are familiar with arithmetic algebra. However, the use of flags often leads to clearer programs, especially for larger projects.

| Exercise 5.12 (MEMO51) |

Draw the function block diagram for each of the three instruction lists. Complete the IL to form complete programs and implement the circuits with the PLC! Make sure that all three programs deliver the same result.

1)		2)		3)	
	LD a		LD a		LD c
	OR b		OR b		AND (a
	ST mem		AND c		OR b
	LD c		ST x		)
	AND mem				ST x
	ST x				

Fig. 5.11 EXOR

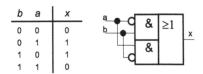

b	a	x
0	0	0
0	1	1
1	0	1
1	1	0

Justify why parentheses are not required in the second variant by tracing the assignments to the "current result". (cf. Exercise 5.11)

5.17   EXOR Operation (Antivalence)

Example 5.9 (Agitator)

An agitator may only be switched on ($x = 1$) when the level is reached ($a = 1$) and the inlet valve is closed ($b = 0$); or when the level is not reached ($a = 0$) and the inlet valve is open ($b = 1$). ◀

You have already dealt with this example for the application of an EXOR element in the digital part in Sect. 2.13 and in the minterms (Sect. 2.14): There we worked out the function plan in detail, so in Fig. 5.11 we only repeat the result.

This EXOR operation is included in IEC 61131-3 and is performed by the XOR operator. The XOR operator can be modified with N and (to XORN and XOR (.

```
Assignment of inputs/outputs (allocation list):
Inputs: Variable a (switch 1) %IX0.6
        Variable b (switch 2) %IX0.7
Output: Variable x (Output x) %QX0.6
```

Exercise 5.13 (EXOR51)

Realize the EXOR link with PLC-lite!

Logical Memory Circuits

6

Abstract

In this chapter you will learn how to use the memories in a PLC. You have already learned about one memory: the flag. In the PLC, a flag is a memory location for a logical value. In the PLC, however, outputs can also be addressed as memory locations. You can assign a value directly to the flags as well as to the outputs in order to reuse it afterwards. You can also set flags to '1' or '0' by a set or reset pulse, similar to flip-flops. In this context, we will have to illuminate special features in the program flow of the PLC, which will then lead us to the function blocks for the RS and SR flip-flops.

In addition to the purpose of recording intermediate results, signals must be stored in order to "record" one-time processes or brief pulses even after they have ended. An application example are circuits in which pushbuttons are used.

6.1 Output with Self-Retaining

Often, processes in systems are triggered by pushbuttons and not by switches. The pushbuttons only make contact as long as they are pressed; when they are released, the signal is no longer present. Two pushbuttons are usually installed: one for switching on and a second pushbutton for switching off. After releasing the ON button, the output must remain at '1' until the OFF button is used to reset the output to '0'. To do this, you need some kind of memory for the button press, because the effect must remain active beyond the duration of the button press. We had seen that this memory behavior can be realized by means of flip-flops.

In electrical engineering, this behaviour can also be achieved by means of an auxiliary contact on a relay. In the circuit diagram for circuit diagram A (Fig. 6.1) you will find a

H.-J. Adam, M. Adam, *PLC Programming in Instruction list according to IEC 61131-3*, https://doi.org/10.1007/978-3-662-65254-1_6

Fig. 6.1 Self-holding: circuit for Exercise 6.1

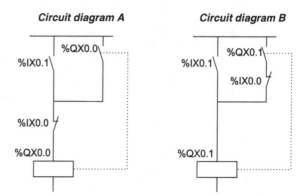

switching contact with the designation %QX0.0. The relay also has the same designation. This means that this switching contact is "actuated" by the relay: when the relay is energised, this contact is closed.

However, this then causes a closed current path for the relay, even if the pushbutton %IX0.1 in parallel with the auxiliary contact is released again, i.e. opened. You can see the memory effect: the relay "holds itself" until the circuit is interrupted by (briefly) opening the pushbutton %IX0.0 in series; then the relay drops out and the auxiliary contact %QX0.0 is opened again.

Exercise 6.1 (POWER61)

Draw the function diagrams for circuit diagrams A and B (Fig. 6.1) respectively, create the instruction list and test the circuits with the PLC. Especially check the behaviour when the two switches %IX0.0 and %IX0.1 are pressed simultaneously.

Note: Note the illustration of the two types of switch: normally open (NO) and normally closed (NC). With the normally open contact, the contact is open in the idle state; actuation closes the contact. In the case of the NC contact, the contact that is closed in the idle state is opened by the actuation. However, this actuation is not important for the PLC, rather the signal emitted by the switch is decisive for the programming, and this is always a '1' when the contact is closed and a '0' when it is open, regardless of whether it is actuated or not!

6.2   Setting and Resetting Outputs

Example 6.1 (Self-Retaining: Start and Stop Agitator with Pushbutton)

An agitator is to be switched on with a start button and switched off with a stop button. ◄

The important thing about this example is that the actions are to be carried out with pushbuttons that jump back to their initial position after being released. You could implement this task with the circuit from Exercise 6.1. In the solution to this example, the agitator at output x is switched on (set) by the start button at b and switched off (reset) by the stop button at a :

Example 6.2 (Self-Retaining)

```
Program POWER61
var
  a AT %IX0.0: bool
  b AT %IX0.1: bool
  x AT %QX0.0: bool
end_var
  LD b
  OR x
  AND a
  ST x
end_program                                                    ◄
```

The circuit works, but admittedly the whole thing is probably too complicated and above all too confusing. Because such a function is needed more often, it is clear that there are special instructions in the PLC for *setting* outputs to '1' and *resetting* them to '0'. In the instruction list, use the operator 'S' to set and the operator 'R' to reset an output or a flag.

Exercise 6.2 (RS61)

Complete the following instructions (Example 6.3) to a complete program and test it with the PLC. The variable a is to be connected to the input %IX0.7, b to %IX0.6 and x to the output %QX0.0.

Example 6.3 (For Exercise 6.2)

```
  LD a
  S x
  LD b
  R x                                                          ◄
```

This program functions as a **self-Retaining** circuit! Pressing the pushbutton a at %IX0.7 sets the output x at %QX0.0 to '1'. This state is maintained even if the pushbutton is released. Only pressing pushbutton b at %IX0.6 clears the output again and resets it to '0'. Note that in each case the action is only executed if the respective variable a or b

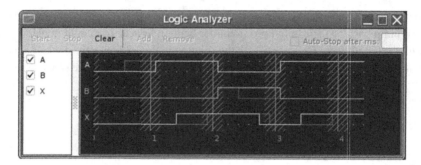

Fig. 6.2 PLC-lite: Logic Analyzer with time diagram for Exercise 6.2

has the value '1'. If the variable has the value '0', no action is executed and the output retains the previous value.

Note
With the help of the Logic Analyzer from PLC-lite you can observe the time sequence of the signals. Open the Logic Analyzer. Then click on "Start" and operate the buttons. You will now get a recording of the logic levels. In addition, you can follow the processes in detail by using the single step mode. In Fig. 6.2 you can see thin and thicker lines in the time diagram. You can understand the differences a little later, with the knowledge gained in Sect. 6.5.

Exercise 6.3

Now observe the behaviour of the output in the program from Exercise 6.2 when both pushbuttons are pressed simultaneously! **In the simulation, you can "clamp" the pushbuttons individually by moving down from the pushbutton while holding down the mouse button.**

6.3   Order of Execution and Priority

From Exercise 6.3 you can see that output x at %QX0.0 always remains reset as long as pushbutton b is pressed, even if pushbutton a is also pressed. Now swap the order of the instructions in Example 6.3 so that the S instruction is the last instruction. You will get Example 6.4:

Example 6.4

```
LD b
R x
LD a
S x                                                                          ◄
```

Now you will see that in Example 6.4 the set command has priority and the output always remains set when both buttons are pressed continuously. We had already achieved such a behaviour with the digital circuits by connecting the signals with switching elements accordingly. Please scroll back to Sect. 3.5.

In the PLC, this arises from the order of the instructions in the instruction list. In the case of multiple assignments to an output within an instruction list, the last assignment "wins". It is therefore of utmost importance in which order the instructions are listed in the instruction list.

> The instructions are executed in order (from top to bottom). In case of multiple assignments, the last one takes *precedence*.

But wait! Please have another look at Sect. 5.10 (Fig. 5.8), which describes how the PLC works: The instructions from the instruction list are not processed only once, but again and again "in a circle" from the first to the last instruction and then starting again with the first. And thus the first instruction is always executed again after the last one. So how can the order of the set and reset instructions matter at all, after all they are simply executed alternately in the end? – We will look at this in the following section:

6.4   Storage of Inputs and Outputs

This behavior: "The last instruction wins" is very useful in both Examples 6.3 and 6.4. It is achieved by the PLC only ever transferring the new values to the output at the end of the cycle, thus giving the output a unique, constant value. The intermediate changes during the cycle remain internal only and do not go to the outside.[1]

---

[1] Likewise, input values are read in only once at the beginning of each cycle and then remain available unchanged until the end of the statement list.

Fig. 6.3 Switching to direct
and indirect access with
PLC-lite

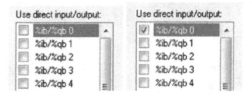

Let us now consider this in more detail: Consider what would happen if the set or reset instructions took effect immediately and directly at the exit, rather than at the end of the cycle?

Exactly! The output would be switched on and off in rapid succession. This would result in a "blinking circuit" with the cycle frequency. You could also say: the output "flutters".

The first instruction resets the output when the button at %IX0.7 is pressed. Immediately afterwards, the output is switched on again when the button at %IX0.6 is also pressed. The best way to study this behavior is to test the program Example 6.2 with the logic analyzer and with the simulator in single step mode (Step).

You have the possibility in PLC-lite via the menu View – PLC-Setup to deactivate the buffering of the terminals, so that the respective values appear directly at the output and not only when the program has run through the cycle to the end. Figure 6.3 shows on the left the default setting for the inputs/outputs: indirect access, with buffering. Toggle: Check the box to set the byte to direct address access. In the following Sect. 6.5, these intermediate memories with the names "process images" are explained in detail.

In Fig. 6.4 the program from Exercise 6.2 is executed in single step mode. PLC-lite marks the line that will be executed next by clicking the "Step" button. In the example, you can see that the execution in "Step" mode is carried out up to line 12. Both input buttons are latched to the value '1'. The output variable x has the value '1' because the previous set command has just been executed. The x value at the output of the PLC (Direct Value) is still '0' because we have *not* selected direct input/output (this is the normal setting). If we now continue the steps, then executing line 13 will reset the output. After executing line 14, this value comes to the output. Despite setting and resetting at the same time, resetting "wins".

Now switch to direct access of the inputs/outputs in PLC-Setup by setting the checkmark. If you now step through to line 12, you will see that the set command has a direct effect on the output and the LED lights up. By executing the reset command in line 13, the output is immediately reset. The LED therefore flashes very briefly during operation. In Fig. 6.5 you can see two completely stepped cycles in the Logic Analyzer. In the third cycle, the set command from line 11 has just been executed.

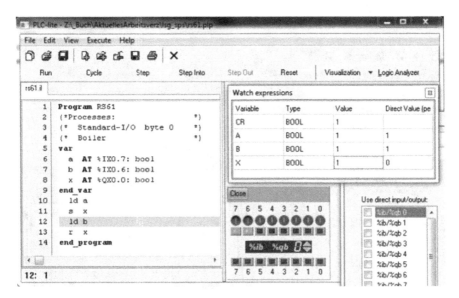

Fig. 6.4 Monitoring variables with PLC-lite in step mode and inputs and outputs in "normal" mode

---

Exercise 6.4

Now run the program from Exercise 6.2 again, now in single step mode and with Logic Analyzer. Observe the behaviour of output x depending on the setting of the inputs "direct input/output" (switched on or off). Also test extensively the case where both inputs have the value '1' at the same time. Now you can also interpret the time diagram of the Logic Analyzer in Fig. 6.2!

In a PLC, the values of the variables within a program cycle must be distinguished from the respective peripheral values, which are read from the process by means of the sensors or which are transferred to the process by means of the actuators. Depending on the application, it may make sense to evaluate the peripheral values directly and immediately in the program or to link them indirectly to the variables in the program by means of the intermediate memories. In addition to the standardized %i and %q, some PLC systems offer additional address ranges or special instructions for this purpose, via which the input/output terminals can be accessed directly.

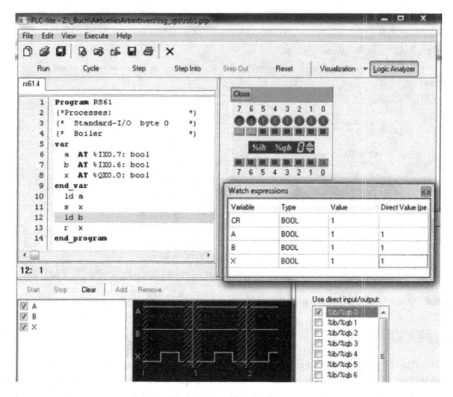

Fig. 6.5 PLC-lite in step mode with Logic Analyzer and inputs and outputs in direct mode

6.5   Process Images of the Inputs and Outputs

The logic results that are assigned to the outputs do not appear immediately at the output terminals, but are first stored temporarily in the "*process image of the outputs*" (PIO or PIQ register, german: PAA). The PIO is only transferred to the output terminals after the last instruction of the instruction list. This ensures uniform, synchronous operation of all outputs.

If, for example, the same output is addressed several times with different values during the cycle, "fluttering" of the output is avoided by using the PIO as a buffer. The assignment to the variable belonging to the output can then be made several times within the cycle; the memory location in the PIO changes its value each time, but only the last assignment in the cycle is actually transferred to the output.

Therefore, in the above Example 6.3, the reset "wins". The program is said to behave R-dominantly. Example 6.4 is primarily setting, S-dominant.

However, the inputs are also not read in directly in the program. Here, the values are read into the *"process image of the inputs"* (PII register, german: PAE) at the start of the cycle. In order to have clear conditions during a cycle, even if an input is queried several times during a program cycle, the input values of all input terminals are only ever read into the "Process image of inputs" (PII) immediately before the start of each program cycle. During the program run, the input values are always read from the PII. Changes made to the inputs in the meantime do not take effect immediately, but only in the next program cycle, after they have been transferred to the PII. This means that the same input can be queried several times in one cycle, and it always has the same value.

As a disadvantage, it should be noted that this means that the shortest, reliably recognizable input pulse must have the length of the cycle duration. Shorter pulses can simply be "swallowed" if they do not last until the next takeover point.

It is often common to refer to the inputs and outputs as "peripherals", as opposed to the variables in the instruction list and process images.

The operation of the PLC is shown in Fig. 6.6. Compared to Fig. 5.8, the process image registers are now also drawn in. This storage of the inputs and outputs in the PII or PIO is carried out by most PLCs. Depending on the automation device, different procedures are implemented to address the inputs or outputs either via intermediate memories or directly as peripheral connections. The latter can be done by allocating certain memory or address areas, by appropriate configuration, by using special commands, etc.

If the PLC does not provide any process images or if they cannot be used in the specific case, we can also take care of the storage ourselves by first applying the set/reset to flags and only transferring the (last) flag state to the output at the end of the cycle. With the help of the flags, we construct our own PIO, so to speak! Correspondingly, we can also use flags to construct a replacement for a PII. You will investigate this in the next task:

Exercise 6.5 (RS62)

Test the two different circuits from Fig. 6.7 with the PLC. Also compare especially the behaviour of the two solutions when both pushbuttons are pressed simultaneously! Observe the assignments to the flags by means of the logic analyzer and in step mode and compare them with the assignments to the outputs! Give reasons for the designations: *priority resetting* and *priority setting*. How can you recognize the respective property in the function block diagram or in the instruction list?

This example shows how the problems of the sequential, cyclic behavior of a PLC can be avoided with a direct assignment even without PII/PIO: by assigning the "intermediate results" to the flag Memo1 or Memo2. Through this indirect assignment, the last executed instruction "wins"; in one case the reset, in the other the set.

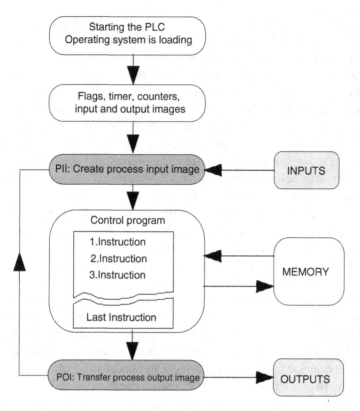

Fig. 6.6 Input image PII and output image PIO (for Exercise 6.5)

Fig. 6.7 Circuits for
Exercise 6.5

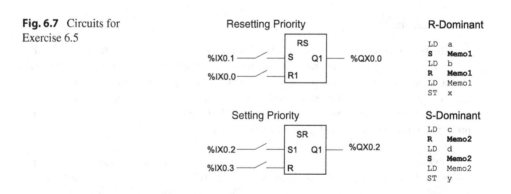

Which of the two circuits in Exercise 6.1 mentioned above is primarily resetting, and which is primarily setting?

6.6   Level Control of a Vessel

The fill level of the vessel in Fig. 6.8 is to be kept between a minimum and a maximum value with the help of a PLC. A button opens valve V3. The container is emptied via valve V3 as long as the button remains pressed.

The control system ensures the required fill level by interrogating the sensors (limit value transmitters) LIS1 and LIS2. The inlet valve V1 is opened as soon as the level drops below the level of LIS1. The valve is only closed again when the level has reached the limit switch LIS2. The signals of the limit switches LIS are '1' when they are immersed in liquid.

```
Assignment list:
LIS1 Lower limit switch   %IX0.1
LIS2 Upper limit switch   %IX0.2
V1 Inlet valve            %QX0.1
V3 Bottom valve           %QX0.3
Sw1 Discharge switch      %IX0.7
```

Example 6.5 (Level)

The level control can be realized with the following program part:

```
LDN LIS1 (* tank empty            *)
S V1     (* then fill (saving!) *)
LD LIS2  (* tank full             *)
R V1     (* then end fill         *)
```

Exercise 6.7 (TANK61)

Create the logic diagram according to the instruction list given in Example 6.5 and program the PLC for the "level control" task. Test the control on the process engineer-

Fig. 6.8 Vessel with sensors
for level monitoring

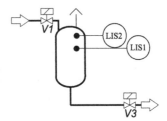

ing model (process "Niveau" or "level"). For emptying, you can press the manual pushbutton at V3. Check whether operating states can occur in which the conditions for setting and resetting occur simultaneously. In this case, the valve V1 could "chatter". Specify which malfunctions in the encoders can lead to impermissible operating states!

Expansion:

With two separate pushbuttons the draining is to be switched on (set) and off (reset) via V3.

6.7   Alarm Circuit 4

An alarm circuit is intended to alert the operating personnel of a plant to dangerous or special operating conditions. In most cases, manual intervention in the process sequence is required. The alarm is triggered by a signal emitted by the process, usually by a horn or/ and a light signal. The "acknowledgement signal", which the plant operator emits by pressing a button, is intended to confirm the alarm. The acknowledgement usually switches off the horn and switches on or over the light signal.

In this application, one-time or short pulses must be detected. Therefore they must be stored. Whether this storage is programmed as priority setting or priority resetting depends on the respective application. If a fault is signalled, the fault signal should be retained as long as the fault has not ended, even if the acknowledgement button is pressed or even blocked in the pressed position. The fault signal must therefore be programmed with priority setting, and thus only after the reset! Accordingly, the acknowledgement signal must be implemented as primarily resetting.

Exercise 6.8 (ALARM61)[2]

Figure 6.9 shows a very simple alarm circuit. The signal S "Fault" is to switch on a horn H (horn), which can only be switched off again by an acknowledgement key "Quitt". Create the instruction list in such a way that, if a fault exists, the horn remains in function despite the acknowledgement key being pressed. The fault signal becomes effective at bit %ix1.1, i.e. in byte 1. In the simulation program, you can open the "Standard I/O" process a second time and switch the byte number by clicking with the mouse on the small digit at the bottom.

```
Assignment list:
Receipt %IX0.6
Fault %IX1.0
```

---

[2] The drawing associated with this exercise (Fig. 6.9) is not complete. It is part of your "homework" to complete the drawing for this and many other exercises.

Fig. 6.9 Circuit for
Exercise 6.8

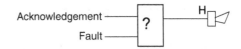

Exercise 6.9 (ALARM62)

A horn is to be switched on by pushbuttons from three different switching points. An acknowledgement button is to switch the horn off again at each of these switching points. Create the circuit with PLC.

Exercise 6.10 (ALARM63)

Extension of Exercise 6.9: It shall be possible to identify by means of three lights from which reporting point the alarm was triggered.

6.8   Signal Memories as Function Blocks

This manual work described in the previous section for the use of flip-flop circuits is not necessary with a PLC according to IEC 61131-3. This is because the flip-flop types are included as so-called *function blocks*. Such a function block independently manages the necessary actions to achieve the priority set or reset.

The priority resetting memory element is called *RS-flip-flop* and the priority setting one is called *SR-flip-flop*. In the graphical representation (Fig. 6.10), the names inside the rectangle are at the top. The names at the inputs and the output within the rectangle are the names of the *"formal parameters"*. The function block uses these names internally. The inputs are on the left and the outputs on the right. The appended digit "1" indicates the priority. Outside the rectangle, you can write the names of the signal variables that you use in your program itself. These are called the *"actual parameters"*.

Fig. 6.10 Function blocks:
priority resetting and priority
setting flip-flop

6.9   Use of Function Modules

The application of function modules takes place in four steps:

1. Define module	*Instantiate*
	Make module available
2. Pass parameters	*Parameterize*
	Supply module with values
	module reads in values
3. Execute module	*Call*
	module works
4. Read out output values	*Results*
	module outputs values
	Values can be used

You can see here the "IPO" model generally known in computer technology: in step 2 the block reads in values (I), in step 3 these are processed (P) and in step 4 the block outputs the results (O).

Step 1: Instantiation
You cannot call the function blocks directly, but you have to create a so-called "instance" for the application. This is much easier than it reads here and goes like assigning the symbolic names to the variable types in the declaration part of the program.

While you specify the data type (e.g. BOOL) for variables, you must specify the function module type to instantiate a function module. This type is the abbreviated name at the top of the graphical display, for example RS or SR.

The effect of instantiation is basically the same for variables and function blocks: for variables, an instance is created which is a memory location of type "BOOL". After instantiation, the function blocks occupy memory space for variables as well as references to functions for the actions that can be performed.

In the example, an instance of a priority-setting memory block is created. You can choose the instance name "FlipFlop" as you like:

```
PROGRAM Pulser
VAR
   FlipFlop: SR
END_VAR
```

Step 2: Parameterization
Before calling the function block in the program (see step 3), you must transfer the values for the inputs to the function block! This is called "parameterizing". You can parameterize in two different ways in the instruction list:

a)   **Call with loading and saving of input parameters**

However, you can also save the actual parameters in the formal parameters before calling them. This has the advantage that the actual parameters are processed *immediately when they are created.*

In the memory commands, the instance name is specified separately with a dot before the formal parameter name. When the block is called later, no more parameters are passed:

```
LD a
ST FlipFlop.R  (* Prepare reset   *)
LD b
ST FlipFlop.S1 (* Prepare setting *)
```

b)   **Call with the list of input parameters**

This method combines steps 2 and 3. When you call it (see step 3), you pass a list with the actual parameters. The operating system then supplies the formal parameters with the current values (see step 2).

In the parenthesis after the instance name of the function block, specify the name of the formal parameter followed by the string ": =". Then the actual value is specified, in this case the identifier of direct PLC inputs. This notation allows you to see the parameter supply "at a glance". *This notation is not possible in PLC-lite.*

```
CAL FlipFlop (R:=a, S1:=b)
```

Step 3: Call in the program

You call the function block in the program using the CAL operator. This function module call executes it, i.e. the output signals are determined from the currently applied input signals and the intermediate values stored in the function module.

```
CAL FlipFlop (* execute function block *)
```

Step 4: Using the output values

You can use the LD operator to read the output values from the function block into the current result and then use them as required, for example, to send them to a PLC output.

```
LD FlipFlop.Q1
ST x
```

Fig. 6.11 Circuit for
Exercise 6.11

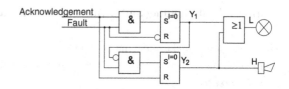

The instruction list must always contain *both* instructions (set *and* reset), because every flip-flop that has been set must also be reset at some point.

Exercise 6.11 (ALARM64)

Examine the alarm circuit 4 (Fig. 6.11). When does the lamp light and when does the siren sound? Also make sure whether the flip-flops should be programmed primarily setting or resetting. Initialisation with the value "0" ("I = 0") is achieved without further action.

Complete the time diagram from Fig. 6.12!

Use the "Boiler" process in PLC-lite. You can generate a fault using the manual button for the heating.

```
Assignment list:
Acknowledge Acknowledge key %IX0.6
Fault Fault signal          %IX1.0
Lamp Indicator lamp         %QX1.0
Horn warning signal         %QX0.7
```

6.10   Control for Filling and Emptying a Measuring Vessel

In chemical process engineering, the individual components must be precisely dosed. Liquids are measured in a vessel. A control system ensures that the vessel is filled precisely. The measuring vessel is always completely filled and completely emptied; this ensures an exact liquid measurement. In the following tasks, you will develop a control system for a measuring vessel.

Fig. 6.12 Time diagram for Exercise 6.11

Fig. 6.13 Measuring vessel
for Exercise 6.12

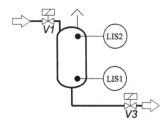

Exercise 6.12 (TANK62)

The measuring vessel (Fig. 6.13) is to be controlled in such a way that water flows in via solenoid valve V1 after an impulse is given via a "Fill" button until the "Full" state is signalled by limit switch LIS2. ("Empty" we program only in the next exercise!) The LIS give signal "1" when immersed in liquid. Create the logic diagram and instruction list. Use the flip-flop function blocks SR or RS for storage! (Process: "Tanks (small)")

```
Assignment list:
Start filling tank (manual switch) %IX0.7
LIS2 Upper limit switch             %IX0.2
V1 Inlet valve                      %QX0.1
```

Exercise 6.13 (TANK63)

Extend the program from Exercise 6.12 so that when a pulse is given via the "Empty" button, the vessel is emptied via valve V3 until LIS1 signals "Empty", i.e. LIS1 supplies a "0" signal.

```
Assignment list (in addition to previous exercise):
S1 Drain container (manual switch)       %IX0.6
LIS1 Lower limit switch                   %IX0.1
V3 Exhaust valve                          %QX0.3
```

Even this solution still shows weaknesses: If you press the Empty button before the vessel is completely full, the water runs through. Now it is possible that neither the upper limit value LIS2 nor the lower limit value LIS1 is reached: the system is blocked!

Exercise 6.14 (TANK64)

Emptying during exercise TANK63 should only be possible when the tank is completely full (LIS2 in liquid). Conversely, filling should also only be possible when the tank is completely empty (LIS1 not in liquid). Test your program thoroughly and also check if during the filling process it is not possible to empty and during the emptying process it is not possible to fill! Correct it if necessary.

At this point we would like to remind you of Sect. 2.7 "AND operation as data switch". You can use the AND operation to switch off data paths and thus prevent inputs from becoming effective. You can use these gate circuits to mutually interlock signals in this task and thus, for example, prevent boiler draining from starting before the boiler is completely filled.

Investigate whether the two memory flip-flops must be priority resetting or priority setting. Draw the logic diagram for this task as well!

With the same arrangement as in the previous Exercise 6.14, the manual button for emptying should now be omitted. The vessel should be emptied automatically as soon as it is full.

Create the associated logic diagram, instruction list and test your program with the PLC and the process engineering process model.

Check how the system behaves when the "Fill" button is held down (blocked in the pressed position) for the entire time!

Try to program a "push-button switch": alternately, pressing *one push-button* (not two push-buttons!) should switch a lamp on and off again. So: the same pushbutton is used both to switch on and to switch off the lamp (as is usual, for example, with bed-side lamps).

   Notice:

This task is difficult to realize with the present knowledge; in Sect. 12.1 a systematic solution possibility is shown by programming as "sequence control". However, it is worthwhile to try this problem now (without "cheating" at the back) in order to see the advantages of the systematic approach all the better later.

Timers with PLC

7

Abstract

In very many processes, an event must last for a very specific time. The controller must then run for a certain time like a stopwatch in response to a start signal. At the end of this time, further actions must be triggered by the program. With a PLC, you can set times from 1/100 s and less to 1 h and more.

In this chapter we will show you how to select and use the appropriate timer types for different tasks in the PLC.

In a PLC there are several different types of timers that can be set and started independently of each other at different times and differ in their response to the start pulse.

In order for the timers to operate independently of each other, they must each have their own memory area for the internal variables. Timers are therefore "function blocks" in the same way as the RS and SR flip-flop signal memories.

7.1   Timer for Pulses

First we consider only the "simple" timer function (Fig. 7.1): the timer has an output which carries '1' signal as long as the time is running and '0' signal when the time has expired. So this timer generates a pulse. According to the standard, its designation is "pulse" or TP ("timer pulse"). It is a function block; it must therefore be instantiated, parameterized and called in a similar way to the flip-flop already discussed.

© The Author(s), under exclusive license to Springer-Verlag GmbH, DE, part of
Springer Nature 2022
H.-J. Adam, M. Adam, *PLC Programming in Instruction list according to IEC
61131-3*, https://doi.org/10.1007/978-3-662-65254-1_7

107

Fig. 7.1 Pulse function block

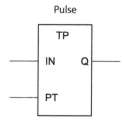

The timer has the input "IN" for start/stop and runs as long as this start/stop input is '1'.

The second input "PT" is used to set the time. Time values (e.g. 100 ms) must be applied to this input. To load the time values into the current result in the PLC program, use the LD command as before.[1]

The instruction is called: LD Value1. For Value1 you can enter t#1000 ms for a time of 1.0 s, for example. For more information on which time values can be used in PLC-lite, refer to Chap. 16 in Table 16.4.

You call the timer with its instance name. The parameterization is carried out as already learned with the flip-flops by the prior assignment of the current parameter values to the formal parameters. You can query output Q with the LD operator. To access the inputs/outputs of the instance, first specify the instance name as usual, followed by a period and then the name of the input or output.

Example 7.1 (Pulses)

Program example for calling the "Timer" function block after loading and saving the input parameters:

```
PROGRAM Pulses
VAR
   Start AT %IX0.0: bool
   Lamp AT %QX0.0: bool
   Pulse : TP
END_VAR
   LD Start
   ST Pulse.IN
   LD t#1000ms
   ST Pulse.PT                                              ◄
```

---

[1] This property that the LD operator can process different data types is called "overloaded". You then simply store the loaded number to the desired location using the ST command, for example to the Impulse.PT input of the timer. The ST operator is also overloaded and can handle multiple data types! See Sect. 8.1 for more information on data types.

Fig. 7.2 Time diagram for
Exercise 7.1 (MIXER71)

```
      CAL Pulse
      LD Pulse.Q
      ST Lamp
END_PROGRAM
```

Have you noticed that a number entered as time is preceded by the character "t#"? You can specify the time unit in any combination of ms, s, m, h, or d (milliseconds, seconds, minutes, hours, or days).

> To use the pulse timer, you must first *instantiate* it. After supplying the parameters, the timer is called by the CAL operator.

Exercise 7.1 (MIXER71)

After pressing a start button, a agitator should run for three seconds. Program the agitator control and test the behaviour. Use the time diagram Fig. 7.2!

Shorten and extend the runtime! How is the timing influenced by the pushbutton? Check in particular the behaviour of the circuit when the pushbutton is released prematurely or pressed several times within the time! Describe in words the processes involved in this control! Note: Use the "Boiler" process! Complete the time diagram (1 graduation mark = 1 s)!

Exercise 7.2 (MIXER72)

There should be an indicator light on while the agitator is running.

Exercise 7.3 (MIXER73)

The agitator in Exercise 7.2 should now continue to run even if the time has expired but the button is still pressed. The lamp should only light up as long as the timer is running. (Do not forget to draw the function chart).

Extend the program so that you can reset the output at any time with a second switch.

Fig. 7.3 Measuring vessel

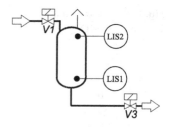

7.2   Filling and Emptying with Time Control

With the control to the measuring vessel in Fig. 7.3, many time-dependent tasks can occur. Perhaps the vessel must not be emptied immediately after filling because the liquid must first settle, or the valve must not be closed immediately after emptying so that, depending on the viscosity of the substance, the vessel can run completely empty.

```
Assignment list:
S1    Drain tank            %IX0.6
S2    Fill measuring vessel %IX0.7
LIS1  Lower limit switch    %IX0.1
LIS2  Upper limit switch    %IX0.2
V1    Inlet valve           %QX0.1
V3    Bottom valve          %QX0.3
Horn  horn                  %QX0.7
```

Exercise 7.4 (TANK71)

The measuring vessel in Fig. 7.3 shall be filled. As soon as it is full, a horn signal should sound for one second. Only after this time has elapsed may emptying be possible.

Exercise 7.5 (TANK72)

After the measuring vessel has been emptied, valve V3 should remain open for another three seconds so that the boiler can run completely empty. Only after this time has elapsed may the filling/emptying process be possible again.

In Exercise 7.5, encoder LIS1 must be interrogated several times. It may happen that the container runs empty exactly between the two queries, so that LIS1 supplies different values in each case. In a PLC, these peripheral values are usually not transferred to the variables directly during the current cycle, but before the start of a new cycle (see Sect. 6.5), so that this case cannot cause a fault.

7.3   Flashing Lights and Oscillating Circuits

With two different timers a circuit can be realized, which oscillates independently. The first timer is started when the second one is not running (anymore); the second one is started when the first one has run out: the timers switch each other.

Example 7.2 (Oscillating Circuit Incorrect)

The following program example seems to do that:

```
program OscillatorERROR
var
  Pulse1:        TP
  Pulse2:        TP
  Lamp AT %QX0.0: bool
end_var
  ld t#1000ms      (* set times *)
  st Pulse1.pt
  st Pulse2.pt
  ldn Pulse2.q     (* is Timer2 running? *)
  st Pulse1.in     (* if no Start Timer1 *)
  ldn Pulse1.q     (* is Timer1 running? *)
  st Pulse2.in     (* if no start Timer2 *)
  cal Pulse1       (* call timer *)
  cal Pulse2
  ld Pulse1.q
  st Lamp
end_program
```
◀

Exercise 7.6 (FLASH71)

Test the program Example 7.2. Find out that it does not work correctly, although the program seems to be correct according to the reasoning from above. To find the error try to find out how the timers start at the very beginning of the program and what happens when a timer runs out!

You can systematically track down the error if you "step through" the program in single steps (Fig. 7.4). In this case, the timers in PLC-lite are controlled so that they run in time with the individual steps. They do not run in real time, but are stopped as long as you leave the program in one step.

Check the times at which the timers are started or updated.

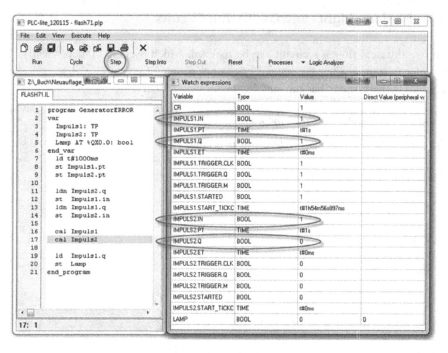

Fig. 7.4 PLC-lite in step mode for Exercise 7.6

In Fig. 7.4 the program OscillatorERROR in PLC-lite is running in single step mode. Each click on "Step" executes another instruction of the instruction list. In the example we have arrived at line 17. The next step would execute timer 2 with `cal pulse2`.

In the "Watch expressions" window you can see that Timer `Pulse1` has already been started and has the output value `IMPULS1.Q` = "1". With the next step Timer `Pulse2` will be started and then its output `IMPULS2.Q` will also = "1". So at the end of the first cycle *both* timers are running: blinking is impossible this way! You see: the timers are always updated when they are called with `cal Timer....` Until the next call, the initial value is maintained. At the beginning of the first cycle *none* of the timers is started, so in the first cycle already *all both* timers get the start signal. The running light cannot work like this!

The solution to this problem is simple: the problem here is that pulse1 is polled *before* it is updated with the `cal pulse1` command. You can work around this problem if you start the timers as soon as the conditions are met, i.e. You keep exactly to the sequence IPO (input – processing – output) and if possible do not insert any other steps in between:

```
ldn pulse2.q
st Pulse1.in (* I: Enter value *)
cal Pulse1 (* P: process in timer *)
ldn Pulse1.q (* O: Output result *)
st Pulse2.in
cal Pulse2
```

Now, when pulse1 is requested (even in the very first cycle!), timer1 is already started, and the program does *not* start timer2 and starts up correctly immediately.

> The timers are updated when they are called with `cal  Timer`. The initial state `Timer.q` is retained until the next call.

The following applies to the duty cycle k and the oscillation frequency f:

$$k = \frac{\text{Duty cycle}}{\text{Total duration}} = \frac{T_1}{T_1 + T_2} \tag{7.1}$$

$$f = \frac{1}{\text{Total duration}} = \frac{1}{T_1 + T_2} \tag{7.2}$$

Exercise 7.7 (FLASH72)

Program the PLC for a flashing light and test the program!

Determine frequency and duty cycle! Change both values, also independently of each other. Also add buttons to start and stop the flasher.

Extend the program to an "alternating blinker", i.e. two lights blink alternately!

7.4   Alarm Circuit 5

Exercise 7.8 (ALARM71)

Analyze the function of the alarm circuit 5 (Fig. 7.5), complete the timing diagram Fig. 7.6 and program the PLC!

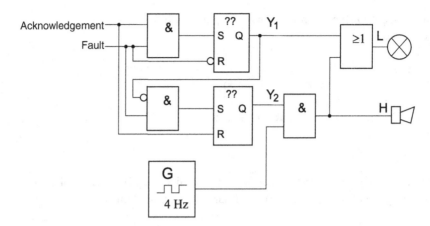

Fig. 7.5 Circuit for Exercises 7.8 and 7.9

Fig. 7.6 Time diagram for
Exercises 7.8 and 7.9

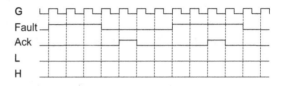

Exercise 7.9 (ALARM72)

Extend the alarm circuit 5 from Exercise 7.8 in such a way that the horn stops immediately after the acknowledgement is pressed, but the lamp continues to flash for another 5 seconds and only then changes to a continuous light if the fault still persists!

7.5   Using Multiple Timers: Chasers

A chaser is a chain of lights that alternately light up one after the other. The circuit for controlling these lights has several outputs; as many as there are lights to be controlled. The circuit can also be thought of as a generator with several outputs, which emits successive pulses on the various outputs.

We want to approach this topic, which at first does not seem so difficult, in steps. First of all, we will build a "mini running light" consisting of only two lamps which are to flash alternately (Fig. 7.7).

Fig. 7.7 Generator with two outputs and example of output pulses (cf. Exercise 7.7)

The extension of Exercise 7.7 to the "alternating indicator" is exactly this task! We go back to this and in Example 7.3 we go straight to the extension to three lights.

Example 7.3 (Running Light with Three Lamps)

For the extension to three lamps, the quick solution is to simply add the third pulse3 to the program from Exercise 7.7:

```
program Flash73ERROR
var
 Pulse1:          TP
 Pulse2:          TP
 Pulse3:          TP
 Lamp1 AT %
 QX0.          0: BOOL
 Lamp2 AT %QX0.1: BOOL
 Lamp3 AT %QX0.2: BOOL
end_var
 ld  t#1s
 st  Pulse1.pt
 st  Pulse2.pt
 st  Pulse3.pt

 ldn Pulse3.q     (* Test Pulse3 *)
 st  Pulse1.in
 cal Pulse1
 ldn Pulse1.q     (* Test Pulse1 *)
 st  Pulse2.in
 cal Pulse2
 ldn Pulse2.q     (* Test Pulse2 *)
 st  Pulse3.in
 cal Pulse3

 ld  Pulse1.q
 st  Lamp1
 ld  Pulse2.q
 st  Lamp2
 ld  Pulse3.q
 st  Lamp3
end_program                                                    ◀
```

But this does not work: When starting a timer you always have to check if *both* other timers are running!

Fig. 7.8 Generator with three
outputs and example of
output pulses

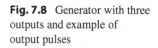

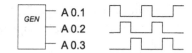

Now create a program for a running light with three LEDs (Fig. 7.8). Perhaps you can
also extend it to more lights?

7.6   Timer with Switch-On Delay

We consider a further timer. In this case, the output signal Q only becomes logic '1' after
a delay time, provided that the input signal is still present. After that, the output remains
'1' as long as the input signal is '1' (Fig. 7.9).

This timing element passes on the input pulse only after a certain delay time has
elapsed. On the other hand, this can also be seen as suppressing (too) short input pulses.

You can study the time sequences with the help of the Logic Analyzer. Figure 7.10
shows an exemplary sequence: The upper line shows the input signal, the lower line the
signal at the output.

Do you recognize that the output signal only appears after a certain time? The output
signal becomes '0' when the input also becomes '0'. If the input signal is too short, no
reaction at all appears at the output.

Fig. 7.9 Timer with switch-
on delay

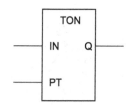

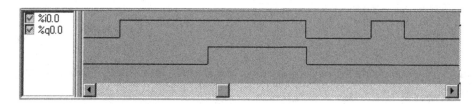

Fig. 7.10 Time diagram in the Logic Analyzer

Exercise 7.11 (FLASH75)

Test the following program Example 7.4. *Attention*, after program start nothing should be visible in the standard process at first. Only when you have held the button at %Q0.0 for at least one second does the lamp start to light up.

Observe how long the output %q0.0 remains logically '1' when you release the key again.

Test this timer extensively to try out other possible applications in the following exercises. Also use the Step Mode and the Logic Analyzer!

Example 7.4 (Switch-On Delay)

```
Program SwitchOn
var
   Pulse: TON
   Start AT %IX0.0: bool
   Lamp AT %QX0.0: bool
end_var
   ld Start
   st Pulse.IN
   ld t#1000ms
   st Pulse.PT
   cal Pulse (* Start timer *)
   ld Pulse.Q
   st Lamp
end_program
```
◀

7.7   Start/Stop Oscillator with only One Timing Element

You can already build a generator, but it has the disadvantage that it requires two timers. In this section we will learn about a generator circuit with only a single timer. The circuit works with the on-delay and a "feedback" from the output to the input.

Look again at the "on-delay" timer. Remember how it works: if a '1' is present at input IN, then output Q becomes '1' at the end of the delay time. If you apply a '0' to the input shortly after this '1' appears, the output also becomes '0' again immediately.

Now we start the game over again and put a '1' at the input again. What happens? Exactly, the output becomes '1' again, but not immediately, but only after the delay time. If now the input becomes '0' again, the '1' at the output is over: there is only a (short) pulse at the output!

Fig. 7.11 Circuit "Start/stop
Oscillator with only one timing
element" for Example 7.5
(pulses)

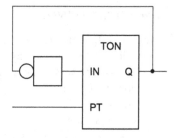

We now make this the method, and always apply the output signal *negated* to the input.
Then we have the circuit shown in Fig. 7.11. The fact that this circuit actually generates
pulses all the time will be examined in more detail in a moment. But first the circuit as
PLC-instruction list:

Example 7.5 (Pulses)

Program code "Start/Stop generator with only one timer": This program emits regular
pulses, but they are of very short duration.

```
Program Pulses
var
 Pulse: TON
 Lamp AT %QX0.0: bool
end_var
 ldn Pulse.Q    (* output negates load and *)
 st Pulse.IN    (* feed back to the input *)
 ld t#500ms
 st Pulse.PT
 cal Pulse      (* Start timer *)
 ld Pulse.Q
 st Lamp
end_program                                                              ◀
```

The Example 7.5 must be analyzed in more detail. For this purpose, we will now
describe the mode of operation step by step. Follow the program run with the aid of
Table 7.1.

We consider the *nth* cycle. This is the last cycle where the output is still '0'. Because of
ldn Pulse.Q there is a '1' at Pulse.IN. As soon as the delay time of the timer has
expired, the output Pulse.Q becomes '1' when cal Pulse is processed. In the following cycle, ldn Pulse.Q is used to apply this signal inverted (i.e. as '0') to input Pulse.
IN. But only when the instruction cal Pulse is reached, the output of the timer will
become '0'.

Table 7.1 Transitions after the delay time has elapsed

	n. Cycle	n + 1. Cycle	n + 2. Cycle
`ldn Pulse.Q`	Output is '0'	Output is '1'	Output is '0'
`st Pulse.IN`	Delay time running	Timer is stopped	Delay time starts again
`cal pulses`	Output becomes '1' when time has elapsed	Output becomes '0'	Output becomes '1' when time has elapsed

The timer therefore had the output at '1' for exactly one cycle! From the next cycle, the delay time starts to run again, because the `Pulse.IN` input now receives a '1' again.

Exercise 7.12 (FLASH76)

Write the sample program 7.5 'Pulse' and test it in step mode.

The pulse duration is exactly a PLC cycle duration. Therefore you will only be able to observe something at output `%q0.0` in step mode or with the help of the logic analyzer.

7.8 Note on the Display of Very Short Pulses

You can make the pulses visible in any case by extending them with a pulse. You can do this with the example program 7.6.

Example 7.6 (ImpulseLang)

This following example program "PulseLong" emits regular pulses whose pulse duration has been extended by means of the timer LED and thus becomes perceptible:

```
Program PulseLong
var
 LED: TP        (* pulse extender *)
 Pulse: TON
 Lamp AT %QX0.0: bool
end_var
 ldn Pulse.Q   (* output negates load and *)
 st Pulse.IN   (* feed back to the input *)
 ld t#500ms
 st Pulse.PT                                    ◀
```

```
cal Pulse       (* Start timer *)
ld t#150ms      (* extend pulse duration *)
st LED.PT
ld Pulse.Q
st LED.IN
cal LED
ld LED.Q        (* output of the LED timer *)
st Lamp
end_program
```

In Sect. 8.7 we will "count" the pulses with a counter and thus make them indirectly visible.

Counter with PLC

8

Abstract

Counting functions are required to implement many control tasks (e.g. quantity or piece count evaluations, evaluation of times, speeds or distances). The counters installed in the automation device can be operated as up or down counters.

In this chapter we will look at counters and data types for numbers.

Up to Chap. 6 we had used exclusively logical datatypes: the values of variables, in- and outputs had only the values '1' ('true') or '0' ('false'). In the program header, these variables were designated with the type 'Bool'. For the type Bool, one memory bit is sufficient, e.g. %IX0.0.

In Chap. 7 "Time functions" another data type was added: times. Times are not logical values, so they are not of type 'Bool'; but we had always specified them directly as constants in the program, but never used them as variables. Since we didn't use time variables, we didn't have to declare them in the program header, so we didn't need to know that they were designated with type 'Time'.

8.1 Data Types

With the counters, however, we cannot avoid using another data type, namely "numbers" also as variables, which must be declared in the program header. We have already dealt with numbers in the first chapter of this book. From Table 1.2 you can see the position values of the dual numbers.

If 8 digits are used, then the largest number that can be represented is:

$$2^7 + 2^6 + 2^5 + 2^4 + 2^3 + 2^2 + 2^1 + 2^0 = 128 + 64 + 32 + 16 + 8 + 4 + 2 + 1 = 255$$

H.-J. Adam, M. Adam, *PLC Programming in Instruction list according to IEC 61131-3*, https://doi.org/10.1007/978-3-662-65254-1_8

The PLC can therefore represent the number range in 0...255 a 8 bit = 1 Byte large memory. This data type is called "unsigned short integer" or "integer" according to IEC 61131: USINT (*unsigned short integer*). We will explain what 'unsigned' means a little later.

These byte values can be declared to %IB0 for the lowest input byte or %QB0 for the lowest output byte:[1]

```
var
  Number8bitIn  AT %IB0: USINT (* with data type *)
  Number8bitOut AT %QB0: USINT
end_var
```

$$01000011_{USINT} = 67_{dez} \tag{8.1}$$
$$10000011_{USINT} = 131_{dez} \tag{8.2}$$

From the pure memory values alone the PLC cannot recognize whether it is a matter of eight individual logical values (bit sequence) or whether the 8 bits mean a USINT number. Therefore, the type specification is indispensable in the declaration.

If a 16 bit = 1 word-sized memory is used, the number range from 0 *to* $(2^{15} + 2^{14} + ... + 2^0) = 65535$ can be represented. The data type is called UINT as abbreviation of 'unsigned integer': These 16 bit values are declared as %IW0 for the lowest input word or %QW0 for the lowest output word:

```
var
  Number16bitIn  AT %IW0: UINT
  Number16bitOut AT %QW0: UINT
end_var
```

The relationship between the memory values for bit, byte and word with the appropriate numbers can be seen in Fig. 8.1. Because two bytes make a word, %IB1 and %IB0 together denote the same inputs as %IW0.

We can not only address a single bit as before (%IXa.b or %QXa.b), but a whole "byte", i.e. 8 bits at once or even a whole "word" with 16 bits! The "lowest" word has the number 0 and is addressed with %IW0 or %QW0. The next higher word is addressed with %IW1 or %QW1.

If you want to represent not only positive but also negative numbers, this is not so easy, because there is no memory for a 'minus' or 'plus' sign: only bits can be represented. As a way out, the following agreement is made: if the first digit is a '0', then the following bits

[1] Note on the two-digit notation for a single bit: The dot is preceded by the byte address; the number following the dot indicates the position of the bit within that byte. Thus, the specification %IX1.0 addresses bit number 0 in the input byte %IB1.

Fig. 8.1 Relationship between bit, byte and word

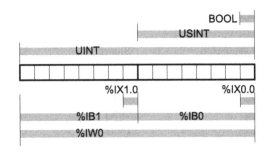

are to be interpreted as positive numbers. But if the number starts with a '1', then it should be a negative number. The first bit serves as a "sign". The representable number range for 8 bits now extends from $-128 \ldots 0 \ldots + 127$. This data type is called a "short integer": SINT.[2]

Also the signed number can be declared with %IB0 for the lowest input byte or %QB0 for the lowest output byte:

```
var
  SignNumber8bitIn AT %IB0: SINT
  SignCount8bitOut AT %QB0: SINT
end_var
```

$$01000011_{\text{SINT}} = 67_{\text{dez}} \tag{8.3}$$

$$10000011_{\text{SINT}} = -126_{\text{dez}} \tag{8.4}$$

The same bit sequence 10000011 resulted in a different decimal value as USINT type! The data types USINT (unsigned number) or SINT (signed number) *can* correspond to different decimal values for the same bit sequence. It is therefore of utmost importance not to confuse the data types.

The signed 16-bit numbers are "integer numbers" of the data type INT, whose value range extends from $-32768 \ldots 0 \ldots + 32767$. The values are declared with %IW0 for the lowest input word or %QW0 for the lowest output word:

```
var
  SignNumber16bitIn AT %IW0: INT
  SignCount16bitOut AT %QW0: INT
end_var
```

Numbers with the length of 32 bits form the data types 'double integer' DINT or 'unsigned double integer' UDINT or 'double word' DWORD. The 64 bit long data types

---

[2] We will not cover in this book how to calculate the negative number from the bit sequence. If you want to know more about this, you can look up the term "two's complement" in e.g. Wikipedia.

Table 8.1 Data types in the PLC

Length (bit)	ANY_NUM		ANY_BIT
	Signed	Unsigned	
1	–	–	BOOL
8	SINT	USINT	BYTE
16	INT	UINT	WORD
32	DINT	UDINT	DWORD
64	LINT	ULINT	LWORD

are: 'long integer' LINT or 'unsigned long integer' ULINT and 'long word' LWORD (cf. Table 8.1).

IEC 61131 summarizes the arithmetic number types in a superordinate type ANY_NUM and the binary bit sequence types in ANY_BIT. Chapter 16, Table 16.5 lists the data types available in PLC-lite.

To load the numerical values into the current result in the PLC program, use the LD instruction as before, regardless of the respective data type. The instruction is called e.g. LD Value1. You have already become acquainted with the property that the LD operator can be overloaded in the previous Chap. 7 (see footnote in Sect. 7.1). In this chapter, USINT values are processed.

In the step mode of PLC-lite you can read the assignment of the different data types with the current values in the window 'Watch Expressions'. In particular, you can track which data types are in the 'Current Result' (CR) in each case.

You then simply save the loaded number to the desired location using the ST instruction, for example to the output %QB1. The instruction is called, for example: ST Out. The ST operator is also overloaded and can process several data types!

Example 8.1

```
Program integer
var
  Value AT %IB0: SINT
  Out AT %QB0: SINT
end_var
  ld Value
  st Out
end_program                                          ◄
```

Exercise 8.1 (INOUT81)

You can use the program from Example 8.1 (Fig. 8.2) to try out reading in and outputting the numbers. Use the other data types as well!
Use both the standard I/O and the hexadecimal digit display (process "HEX Output"). Compare the numbers in the dual with the hexadecimal display.

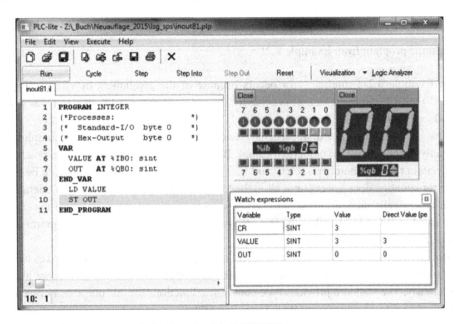

Fig. 8.2 PLC-lite in step mode for Exercise 8.1 (INOUT81)

You can load all processes at the same time. By clicking on the small digit, you can switch the byte number via which the process is to be "connected" to the PLC. Adjust the byte number accordingly in the program!

8.2 Type Conversions

Example 8.2 (Assignment of Different Data Types)

```
Program integer
var
  Value AT %IB0: USINT
  Out AT %QB0: SINT
end_var
  ld Value
  st Out
end_program
```
◄

Test the program from Example 8.2!

In Example 8.2 a runtime error occurs because of the assignment of Value (type: USINT) to Out (type: SINT). This behavior did not exist in the "old" PLCs, it was only introduced with IEC 61131.

This is a bit annoying at first, especially since the 8.2 program would work completely correctly depending on the numerical values. But it turns out to be a great help in practice: In a certain range of numbers, SINT and USINT are the same. It could therefore be that the error situation does not occur at all in the test phase (by chance), but later in operation other numerical values occur, which then lead to a functional error. The type checking performed in IEC 61131 makes the PLC programs safer by detecting such programming errors in good time.

But there are situations conceivable, where bit sequences have to be considered as numbers of different types. However, the programmer knows this and can also carry it out consciously. The standard provides type conversion functions for this purpose. For example USINT_TO_SINT in Example 8.3.

These type conversions do not take place automatically in a PLC according to IEC 61131, but only after explicit programming. This clearly places the responsibility on the programmer to ensure that the conversion does not cause any errors in the specific case. We will return to this issue later in Exercise 8.4 and in Exercise 8.8.

Example 8.3 (Assignment after Conversion of the Data Type)

```
Program IntegerType
var
  Value AT %IB0: USINT
  Out AT %QB0: SINT
end_var
  ld Value
  USINT_TO_SINT (* Data type conversion *)
  st Out
end_program                                                 ◄
```

Exercise 8.3 (INOUT83)

Test the program Example 8.3 for different input values, especially also for values >127. In single-step mode, observe the assignment of the data types to the Current result CR!

Fig. 8.3 Up-counter CTU

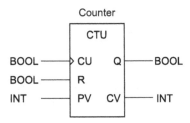

8.3 Three Different Counter Types

8.3.1 Up Counter

In Fig. 8.3 the current counter value is available at output CV as a value with the number type INT. The identifier CV means "counter value". How we handle this number further, you will learn a little later in this chapter.

The counters in the PLC respond to pulses on the CU input and change the count value at the start of the pulse when the input value goes from '0' to '1'. This is called: rising edge triggering. The second, falling edge at the end of the pulse has no effect on the count result.

The "up-counter" increases its numerical value with each pulse (more precisely: with each rising edge) at the CU input, it is a "counter up". The abbreviation in the circuit diagram is therefore "CTU".

The CTU counter has three inputs, designated by the formal parameters CU, R and PV. The CU input is for the pulses to be counted. CU stands for "clock up". The R input is used to *reset* the counter value to zero. A '1' at the R input immediately sets the counter value to zero.

Step 1: Set Preset Value
The PV input requires the number type INT. Assign a number to the formal parameter PV using the ST operator. This gives the counter a comparison number to control output Q. The abbreviation PV means "Preset Value".

Step 2: Evaluation of the Counter
You can query the Q output to evaluate the counter. This is an output with a Boolean value. It is only '1' if the counter reading at the CV output is greater than or equal to the preset value at the PV input, otherwise it is '0'.

Notice: When the PLC is started, the value PV is preset to zero.

The output Q of the up-counter is '1' if the current count value CV is equal to or greater than the preset value set at PV.

Step 3: Output of the Numerical Values
The numerical values of the current counter status at output CV are available as integer numerical values (16 bit) binary coded.

Exercise 8.4 (COUNT81)

Test the up-counter with the following sample program "CountUP". In particular, explore when output Q changes its signal (process "Counter CTU")!

For practice, try to convert the dual displayed numerical values into hexadecimal numbers. Use the HEX display only for checking! Refer to the digital technology section at the front of the manual if you have problems with the dual and HEX display.

With PLC-lite you can also use the step mode here. Please note that you must not simply "clamp" the clock input CU, but that you must apply at least one cycle alternately '1' and '0' (i.e. a complete pulse). (see Figs. 8.4 and 8.5)

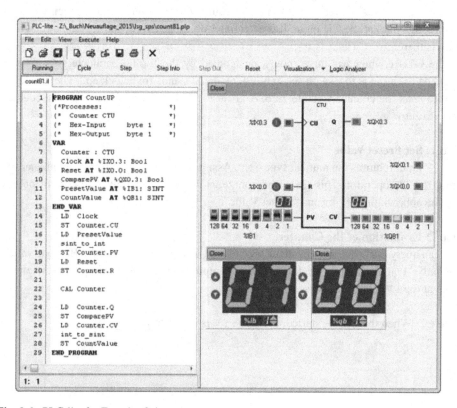

Fig. 8.4 PLC-lite for Exercise 8.4

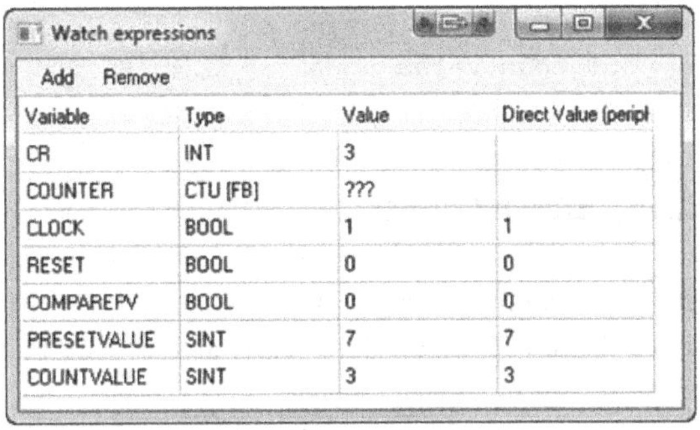

Watch expressions

Add Remove

Variable	Type	Value	Direct Value (peripl
CR	INT	3	
COUNTER	CTU [FB]	???	
CLOCK	BOOL	1	1
RESET	BOOL	0	0
COMPAREPV	BOOL	0	0
PRESETVALUE	SINT	7	7
COUNTVALUE	SINT	3	3

Fig. 8.5 PLC-lite single step for Exercise 8.4

If you take a closer look at the example program, you will notice that the explicit type conversion described in Sect. 8.2 is used here: this is necessary because the counter function block works internally with `INT` variables, i.e. 16 bits, while only 8 bits each (data type `SINT`) are available for the input and output in the example.

Consider under which conditions this type conversion is valid! Check your consideration with the simulation by provoking a runtime error there. To prevent these errors, you must check the value range before the type conversion and take appropriate action in the event of an overflow. You will learn an example of this in Sect. 10.5.

Example 8.4

```
PROGRAM CountUP
VAR
  Counter :    CTU
  Clock        AT %IX0.3: Bool
  Reset        AT %IX0.0: Bool
  ComparePV    AT %QX0.3: Bool
  PresetValue AT %IB1: SINT
   CountValue  AT %QB1: SINT
END_VAR
  LD           Clock
  ST           Counter.CU
  LD           PresetValue
  SINT_TO_INT                  (* type conversion! *)
  ST           Counter.PV
  LD           reset
  ST Counter.R
  CAL Counter (* Call counter *)
  LD Counter.Q
  ST ComparePV
  LD Counter.CV
```

```
        INT_TO_SINT (* type conversion! *)
        ST CountValue
    END_PROGRAM                                                      ◄
```

8.3.2 Down Counter

The down counter in Fig. 8.6 has the name: CTD ("Counter *Down*"). It reacts to the rising edges at the CD input. With each active edge at the CD input, the counter value CV decreases by 1.

A '1' at the LD input sets the counter value CV to the preset value PV.

Attention

The preset value is always 0 when the PLC is started. The down counter thus counts into the range of negative numbers from the first pulse. If you want to process the counter reading as an unsigned number (e.g. as USINT), you must therefore *always* load a (positive) preset value yourself, otherwise you cannot count down!

Also with the CTD counter you can make a quick evaluation of the counter. However, because this counter counts backwards, it is interesting to know whether a lower value has been reached or fallen below. This lower value cannot be changed and is always zero.

> The output Q of the down counter has the value '1' if the counter reading CV is less than or equal to zero.

Exercise 8.5 (COUNT82)

Create a program to test the possibilities of the down counter! Pay attention again to the output value CV of the counter and the signal Q (process "Counter CTD")!

8.3.3 Combined Up/Down Counter

There is also the combined counter type, which has two separately operating pulse inputs (Fig. 8.7). The counter value CV registers the pulses at the CU and CD inputs independently of each other. The meanings of the inputs and outputs are the same as those of the other

Fig. 8.6 Down counter CTD

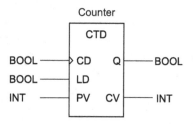

Fig. 8.7 Combined up/down
counter CTUD

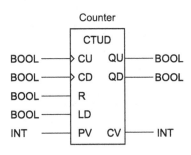

counter types. However, there are now two outputs which indicate whether the counter
value CV has exceeded the preset value PV or fallen below '0'. They are called QU and QD
accordingly.

Exercise 8.6 (COUNT83)

Create the sample program and test the function (process: "Counter CTUD"). At which
clock edge does a counter pulse become effective? When do the outputs QU and QD
become '1'?

Example 8.5

```
PROGRAM CountUPDN
VAR
  Counter : CTUD
  ClockDN      AT %IX0.2: Bool
  ClockUP      AT %IX0.3: Bool
  Reset        AT %IX0.0: Bool
  SetLoad      AT %IX0.1: Bool
  CompareZero AT %QX0.2: Bool
  ComparePV   AT %QX0.3: Bool
  PresetValue AT %IB1:    SINT
  CountValue  AT %QB1:    SINT
END_VAR
  LD PresetValue
  SINT_TO_INT      (* type conversion! *)
  ST Counter.PV
  LD ClockDN
  ST Counter.CD
  LD ClockUP
  ST Counter.CU
  LD reset
  ST Counter.R
  LD SetLoad
  ST Counter.LD
```

```
    CAL Counter
    LD Counter.QU
    ST ComparePV
    LD Counter.QD
    ST CompareZero
    LD Counter.CV
    INT_TO_SINT      (* type conversion! *)
    ST CountValue
END_PROGRAM                                              ◄
```

8.4   Determine Numbers

When a certain number is reached, a Boolean value is to go to '1' and thus trigger an action. The Q output of the counter can be used for this.

Exercise 8.7 (COUNT84)

In a retail store, a control light in the office should light up when one or more customers are in the store. Design a solution with PLC.

In addition, you can also output the total number of customers.

As a further modification, you can change the output so that lamps indicate whether there are no customers or more than 5 customers in the store.

8.5   Multi-Digit Decimal Counter (BCD)

We now want to implement a multi-digit decimal counter. You already learned about such a counter in Part I of this book. Go back to the BCD counter (Sect. 4.6.5), and to Sect. 1.12 for the BCD code.

Exercise 8.8 (DEKADE81)

Complete the following program and add to it a three digit output. Perhaps you can also create a reset option to reset the decades to zero by means of a pushbutton, independent of the respective counter reading. To visualize the output you can use a "Hex-Output" for each decimal place.

Justify why type conversion with INT_TO_SINT may be used in this particular case without checking the value range separately.

Example 8.6

```
PROGRAM Decade81
VAR
    One : CTU
```

```
Ten : CTU
Clock AT %IX0.3: Bool
ByteOne AT %QB1: SINT
ByteTen AT %QB2: SINT

END_VAR
   LD 10        (* default setting *)
   ST One.PV
   ST Ten.PV
   LD Clock     (* Clock            *)
   ST One.CU

   CAL One      (* Start single digit           *)
   LD One.CV    (* Read value                    *)
   INT_TO_SINT  (* type conversion OK here! WHY? *)
   ST ByteOne   (* output                        *)
   LD One.Q     (* Test overflow                 *)
   ST Ten.CU    (* input pulse higher digit      *)
   ST One.R     (* Reset position                *)
...
END_PROGRAM                                                  ◀
```

8.6   Multiple Filling and Emptying

In a chemical process, the individual components must be in a certain ratio. The dosing is done by the number of fillings of the measuring vessel. Of course, for different substances to be mixed in the reaction vessel, either several measuring vessels are needed, or different substances are filled into the measuring vessel through different inlet valves. In a first example we will use only one measuring vessel and also only one inlet valve.

Exercise 8.9 (MIXER81)

The control is to be programmed in such a way that the measuring vessel is filled via V1 after an impulse is given via a start button and emptied via V3 when LIS2 has responded. This process should run three times in succession. After the third emptying, it should be possible to restart the entire process. In this task, do not yet consider the processes in the reaction vessel!

After the assignment list, create the function block diagram and the instruction list for the PLC. Test the program with the model from Fig. 8.8.

```
Assignment list:
S1   Fill measuring vessel/ start control      %IX0.7
LIS1 Lower limit switch                        %IX0.1
```

Fig. 8.8 Process model

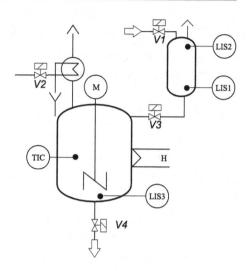

```
LIS2 Upper limit switch                                %IX0.2
LIS3 Limit value transmitter Reaction vessel %IX0.3
TIC  Setpoint generator temperature                    %IX0.4
V1   Inlet valve                                        %QX0.1
V3   Exhaust valve                                      %QX0.3
V2   Valve coolant                                      %QX0.2
V4   Outlet valve reaction vessel                       %QX0.4
M    Stirrer motor                                      %QX0.6
H    Heating                                            %QX0.5
```

8.7   Generator for Counting Pulses

We now return to the task from Sect. 7.7 (single timer start/stop generator). This generator produces pulses of the duration of a single cycle. These short pulses can be applied to the input of a counter. The counter run will tell you if the pulses are present, even without a logic analyzer.

Exercise 8.10 (GEN81)

Complete the program Example 7.5 "ImpulseTON" with a counter and the corresponding display (Fig. 8.9).
You can also use the multilevel decade counter from Exercise 8.8 (DEKADE81).
  Change the count speed. Pay close attention to how you change the values to make the counter run faster or slower.

Fig. 8.9 Circuit for Exercise
8.10 (GEN81)

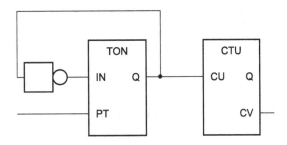

Start/Stop Device

To set up a start/stop facility, remember the gate circuit in Sect. 2.7. Make sure that the "feedback" of the `Pulse.Q` signal to the `Pulse.IN` input can be interrupted.

Exercise 8.11 (GEN82)

With the switch at `%i0.7` the count run shall be started and stopped (use also the process "HEX-Output").

Exercise 8.12 (GEN83)

Now use a combined up/down counter, connect a stopable pulse generator to each of the counter inputs `CU` and `CD` and observe the counter readings. Complete the circuit by loading a preset value. The counter should only be able to count back and forth between zero and this maximum value!

8.8   Time measuring

Basically, time measurement means nothing more than counting: a generator forms the "time standard", which emits pulses at regular intervals. If the pulse intervals are exactly one second, then the counter reading indicates the number of seconds that have elapsed. With shorter pulses, e.g. 1/100 s, correspondingly more precise times can be realized. The single timer start/stop generator shown in Sect. 7.7 is ideal for timing.

Exercise 8.13 (TIME81)

Start the counter with the button at `%i0.1` and determine the time until you stop the counter again with the button at `%i0.2`. With the key at `%i0.6` or `%i0.0` the counter is to be reset again.

Exercise 8.14 (TIME82)

Determine the time in which the measuring vessel is filled from the lower level LIS1 to the upper level LIS2.

8.9   Quantity Measuring

The time for filling the measuring vessel is always the same if the inflow is constant. If you measure the filling time, you can also interpret the meter output as the filling quantity. The measuring vessel content would be exactly 50 l between the two levels. You can follow the litre display in the following task and stop "by hand" at a certain number of litres. In Chap. 10 we will get to know a possibility to let the inflow stop automatically when a certain number is reached.

Exercise 8.15 (TANK81)

Set the pulse duration so that the counter runs from 0 to 50 during filling from LIS1 to LIS2. Attention: if you use the HEX output, the numerical value runs up to the HEX number (32_{Hex} or as binary value 00110010), which corresponds to the decimal value 50_{Dez}.

A decimal display instead of hexadecimals would be desirable here. Unfortunately, you will have to wait a while for this: We will create a converter for HEX to BCD numbers later, in Sect. 11.4.

8.10   Reaction Tester

In this section, you are to solve a task in which two timers are used. At the push of a button, a timer starts, which determines a lead time. After its expiration, a lamp lights up and the pulsetimer starts to run.

Its pulses are counted until it is stopped by pressing a button. This makes it possible to determine the reaction time that elapses between detecting the light signal and pressing the button.

In this exercise, we set a specific lead time. In Sect. 10.9 we will learn to generate random numbers.

Exercise 8.16 (REAKT81)

Program a reaction tester!

Function Blocks

9

Abstract

You already know some of the standard function blocks present in every PLC: RS and SR flip-flops, timers and counters. In this chapter, we will look at creating your own function blocks. You will learn how to create function blocks yourself. Function blocks are used as a tool for structuring PLC programs. You can program frequently used or complicated actions once in self-created function blocks and then use them as often as you like. These self-created function blocks are called and used in exactly the same way as the already familiar standard function blocks: namely instantiated, supplied with parameters, called and the results retrieved.

A function block is a program organizational unit that returns one or more values during execution. Several instances of a function block can be created, each with its own name and memory areas for the variables.

Any function block that has already been declared can be used in the declaration of another function block.

9.1 The Function Block (FB) Outputs Values

Example 9.1 (Flashing by Means of a Function Block)

If a '1' is present at input %IX0.1, output %QX0.0 is to flash. You have already carried out this task in Exercise 7.6. Now, however, this program is to be rewritten as a function block so that it is easier to integrate the blinking function into a program. ◄

© The Author(s), under exclusive license to Springer-Verlag GmbH, DE, part of
Springer Nature 2022
H.-J. Adam, M. Adam, *PLC Programming in Instruction list according to IEC 61131-3*, https://doi.org/10.1007/978-3-662-65254-1_9

9.2   Creating a Function Block (FB)

- The entire code of a function block is framed by the keywords FUNCTION_BLOCK
  and END_FUNCTION_BLOCK.
- Variables that are only used within the function block (local variables) are declared
  between VAR and END_VAR. In the example, the timers, which are only required
  within the function block, must be instantiated here. They are then available as local
  values in the function block.
- Variables that are used for output from the function block to the calling program must
  be declared between VAR_OUTPUT and END_VAR. In this way, they are assigned their
  value in the function block, which you can load ("read") in the main program. Note that
  'Q' cannot be used as the output name, because the identifier 'Q' is reserved for the
  outputs of the standard function blocks![1]
- The actual program code then does not differ from a "normal" program, e.g. the following function block Gen05 is practically identical to the program already written in
  Exercise 7.6.
- The end of the block is indicated by RET. This means that the program execution is to
  return to the calling (main) program.

Example 9.2 (Function Block for an Oscillator)

```
FUNCTION_BLOCK Gen05
VAR
   Pulse1 : TP
   Pulse2 : TP
END_VAR
VAR_OUTPUT
   Y : BOOL
END_VAR
   LD t#500ms
   ST Pulse1.PT
   ST Pulse2.PT

   LDN Pulse2.Q
   ST Pulse1.IN
   CAL Pulse1

   LDN Pulse1.Q
   ST Pulse2.IN
   CAL Pulse2
```

---

[1] For reserved identifiers in PLC-lite, see Sect. 16.3 (Table 16.12).

```
LD Pulse1.Q
ST Y
RET
END_FUNCTION_BLOCK
```
◀

Exercise 9.1

Compare the function block `Gen05` from Example 9.2 with the program from Exercise 7.6 and mark the changes in the `FUNCTION_BLOCK Gen05` block by underlining them.

9.3   Program Organization Units (POUs)

Example 9.3 ("Main Program", Which Uses FB Gen05)

```
PROGRAM FlashLight
VAR
   Flash: Gen05              (* FB declaration *)
   Lamp AT %QX0.0: Bool
END_VAR
   CAL Flash
   LD Flash.Y
   ST Lamp
END_PROGRAM
```
◀

The function block `Gen05` must first be instantiated (like a standard function block) in the program header of the main program and given the instance name.

The call is made in the program body of the "main program" with the instruction `cal Flash`. After that, the output of the flasher can be queried. The necessary timers are "hidden" for the program in the function block `Gen05`.

Exercise 9.2 (FLASH92)

Create the program for the turn signal as described above.

Type the Example 9.2 and save the *file* under the name `FB_GEN05.IL`. *Attention!* Do not confuse this with saving the *project*!

In order to add this main program to the project, select in PLC-lite in the menu File the item: New file, or click the symbol for "Add new file" (Fig. 9.1). This will create a new sheet in the editor where you can write the main program. (Fig. 9.2) You can switch between the individual files using the tabs at the top of the editor window. At the end, save this *file as* well. Use the name `FLASH92.IL` for the main program.

Fig. 9.1 Creating a new file and adding it to the project

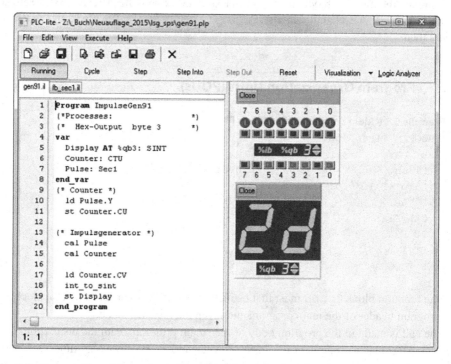

Fig. 9.2 Program files of the project of Exercise 9.3 and simulation in PLC-lite

The entire project now consists of two *program files*: one with the main program and the other with the function block.

Each of these parts is called a *"program organization unit"* (*POU*). In the PLC system, the project management ensures that the parts that belong together are also found together again. In PLC-lite, you only have to call up the "Save project" item in the File menu. The next time the project is loaded, all associated files are opened again and displayed in the editor.

Note the names: The POU "Main Program" has the name you entered after the keyword Program. This name does not have to be identical with the name of the file on the hard disk!

The same applies to the POU "function block": The name of the function block is located after the keyword Function_Block. The function block is addressed under this name in the main program when it is instantiated and called. On the hard disk, the function module can be given a different name when it is saved.

Exercise 9.3 (GEN91)

Create a function block FB_SEC1 which, as in Exercise 8.10, generates pulses at 1-second intervals. Test the function block in a program with an up-counter.

9.4   Including a Function Block in a Project at a Later Stage

You can try out the advantage of these function blocks right away: an already written program is to be extended by the possibility of a blink output. Instead of having to rewrite everything in this program, you can simply include the already existing function block in the project! (Fig. 9.3).

Exercise 9.4 (ALARM91)

Load the project ALARM61.PLP. This initially contains only the file ALARM61. IL. Save the file under the name ALARM91.IL. Add the file FB_GEN05.IL and save the *project* under the name ALARM91.PLP. After instantiation of the "Blinker" you can modify the program so that in addition to the alarm horn the indicator lamp flashes at %QX0.0.

Fig. 9.3 Including an existing file in an open project

Fig. 9.4 RS flip-flop made
from standard components.
Converted from the circuit in
Sect. 3.2

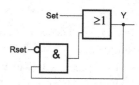

9.5   The Function Block Reads in Values

In Sect. 3.2 we created a flip-flop. In the circuit of that time only the *inverted* flip-flop
output is present. By conversion with the help of de Morgan's rules we get the circuit as
we have given it to you here in Fig. 9.4.

Exercise 9.5

Try to prove the equality of the circuit in Fig. 9.4 with the circuit you already know
from Sect. 3.2!

We now want to create this circuit as a function block. The identifiers for the block
name (RS or SR) and the names of the formal parameters (R, S, Q) already provided by the
standard function block are protected keywords that must not be used in your own projects. Let us use the identifiers Set and Rset for the inputs and Y for the output. The block
is to be called FFSR.

Now values have to be brought into the block as well as out of it, we get INPUT and
OUTPUT variable. Here the beginner often has problems with the assignment. The values
that the *main program* supplies are *input values* for the function *block*. The values that the
block supplies are its *output values,* but they are input values for the calling main program!

So that no confusion arises here:
Input and output are always to be seen from the *module*!

Example 9.4

```
FUNCTION_BLOCK FFSR
VAR_INPUT
   Set  : BOOL
   Rset : BOOL
END_VAR
VAR_OUTPUT
   Y :    BOOL
END_VAR
```

Fig. 9.5 Circuit for
Exercise 9.7

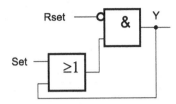

```
LD    Set
ORN( Rset
AND  Y
)
ST   Y
RET
END_FUNCTION_BLOCK                                                          ◀
```

Exercise 9.6 (FFFB92)

Open a new project and write the function block FFSR according to Example 9.4. Save
the function block under the name FB_FFSR.IL. Then write a main program that
uses this block. The inputs %IX0.0 and %IX0.1 are to be reset and set inputs respectively, the output %QX0.0 is to display the result. Explain why the function block is
primarily setting.

Exercise 9.7 (FFFB93)

Write a function block FFRS (file name: FB_FFRS.IL) for a priority resetting flip-
flop. The standard specifies the circuit shown in Fig. 9.5 for this type.

9.6   Function Block: FB_Tank

Example 9.5 (Tank Function Block)

A function block FB_Tank is programmed according to Exercise 6.15 and provided as
a function block with the input parameters: LIS1, LIS2 and Start.

Start = '1' triggers a filling process of the boiler. The function block controls the
filling or empty valve via the output parameters V1 and V3.

```
FUNCTION_BLOCK Tank
VAR_INPUT
  LIS1 : BOOL
  LIS2 : BOOL
```

```
   Start: BOOL
END_VAR
VAR_OUTPUT
  V1 :     BOOL
  V3 :     BOOL
END_VAR
VAR
  V1FF :  RS
  V3FF :  RS
END_VAR
  LD    Start
  ANDN V3
  ANDN LIS2
  ST    V1FF.S
  LD    LIS2
  ST    V1FF.R1
  ST    V3FF.S
  LDN   LIS1
  ST    V3FF.R1
  cal   V1FF
  cal   V3FF
  ld    V3FF.Q1
  st    V3
  ld    V1FF.Q1
  st    V1
END_FUNCTION_BLOCK                                ◄
```

Example 9.6 (OneTank Main Program)

The main program consists almost only of parameter calls. You don't notice anymore
that flip-flops are required!

```
program OneTank
var
(* instantiation *)
  Tank0: Tank
  LIS01  AT %IX0.1: bool
  LIS02  AT %IX0.2: bool
  V01    AT %QX0.1: bool
  V03    AT %QX0.3: bool
  Start  AT %IX0.7: bool
end_var
(* Linking the input terminals *)
```

```
    ld Start
    st Tank0.Start
    ld LIS01
    st Tank0.LIS1
    ld LIS02
    st Tank0.LIS2
(* function module call *)
    cal Tank0
(* Linking the output terminals *)
    ld Tank0.V1
    st V01
    ld Tank0.V3
    st V03
end_program                                                          ◄
```

Exercise 9.8 (TANK91)

Write down the function block (Example 9.5) and the main program (Example 9.6). Save the function block under the name FB_Tank.il! The program file should be named TANK91.IL and the project TANK91.PLP. Test the project with the process "Tanks (small)"!

So far, I don't think you see much of an advantage. The story becomes interesting when several boilers are used. Then the program no longer needs to be rewritten for each individual boiler, but it is sufficient to instantiate the function block for each real boiler and link the input and output terminals with the formal parameters.

Exercise 9.9 (TANK92)

Complete the program from Exercise 9.8 so that all three boilers are filled at the push of a button, and test it with the "Tanks (big)" process! You can find the beginning in Example 9.7.

Example 9.7 (Extension of Exercise 9.8 to a System with Three Tanks)

Note that the *function block* FB_Tank does not have to be changed here compared to Exercise 9.8! You do not need to retype it or save it again.

```
program ThreeTanks
var
(* instantiation tank0 *)
    Tank0: Tank
    LIS01 AT %IX0.1: bool
    LIS02 AT %IX0.2: bool
```

```
  V01   AT %QX0.1: bool
  V03 AT %QX0.3: bool
(* instantiation Tank1 *)
  Tank1: Tank
  LIS11 AT %IX1.1: bool
  LIS12 AT %IX1.2: bool
  V11   AT %QX1.1: bool
  V13   AT %QX1.3: bool
(* instantiation Tank2 *)
  Tank2: Tank
  LIS21 AT %IX2.1: bool
  LIS22 AT %IX2.2: bool
  V21   AT %QX2.1: bool
  V23   AT %QX2.3: bool
  ...
```

Jumps, Loops and Repetitions

10

Abstract

For *execution control*, conditional and unconditional *jump instructions* are available in the "Instruction List" (IL) language. With their help, repetition structures ("loops") and selections can be realized. This allows us to program complex sequences that either repeat the same statements several times in succession or whose sequence depends on input values or other conditions and can take different "paths" in the statement list.

The language "Structured Text" (abbreviated ST) goes even further, but is not the subject of this course. It is a high-level programming language for PLCs according to IEC 61131-3, which allows to solve repetition statements or branches with the constructs FOR ... TO ... DO, WHILE ... DO, REPEAT, CASE, IF ... THEN ... ELSE very elegantly as known from other high-level languages.

10.1 The Running Dot

We have already dealt with running lights in Sect. 7.5. Unlike there, only *one* timer is to be used here, which is operated as a pulse generator and supplies switching pulses. With each pulse a new *numerical value* must be applied to the output.

How can the light spot continue to run? Go back to Sect. 1.6 and assume that the LED at the rightmost byte %qb1 is lit. The value of this byte in binary notation is: 0000 0001. In the decimal system, this is the value 1. After the switching pulse, the light spot should have moved one position to the left; the binary value is then: 0000 0010, which corresponds to 2 in the decimal system. One step further, the value 0000 0100 ($=4_{\text{decimal}}$) must

© The Author(s), under exclusive license to Springer-Verlag GmbH, DE, part of Springer Nature 2022
H.-J. Adam, M. Adam, *PLC Programming in Instruction list according to IEC 61131-3*, https://doi.org/10.1007/978-3-662-65254-1_10

be output. Now the calculation is clear? Exactly, the numerical value must be doubled when moving on. The operation `mul  2` multiplies the current result by 2.

> The operation `mul  2` multiplies the current result by `2.` In the case of the binary number, this means the shift by one place to the left.

10.2   Conditional Jump

We use the pulse generator from Sect. 7.7.

The switching on of the chaser must take place exactly at certain points in time. We therefore need the condition for switching on at these time intervals. The PLC always runs through cyclically; this means that the multiply instruction may only be executed when the "correct" time has come. Then the `mul` instruction must be executed in the cycle, otherwise not. This can be achieved by skipping the `mul 2` instruction during the "wrong" times.

Our pulse oscillator supplies a '1' signal for exactly one cycle after the delay time. Thus the switching condition can be formulated relatively simply (Fig. 10.1): as long as the oscillator delivers a '0', the command `mul  2` must be skipped. Exactly in the cycle in which the pulse oscillator delivers the '1', the new numerical value must be generated, i.e. the `mul  2` command must be executed.

The mechanism for this skipping consists of two parts:

- firstly, an *evaluation of the condition* whether to jump or not,
- and secondly the *target*, i.e. the point in the program at which machining is to be continued in the event of a jump.

Fig. 10.1 Flow chart "conditional jump": the jump is executed if Q = 0 is, otherwise the instruction in the box (mul 2) is executed

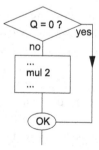

Fig. 10.2 Structure chart for
the conditional jump as in
Example 10.1

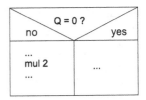

Example 10.1 (Conditional Jump)

```
ldn Pulse.Q
jmpc OK    (* Test of the jump condition *)
   ...     (* part to be skipped         *)
OK:        (* Jump target mark:          *)
```
◀

The jump condition is tested with the command `jmpc`. The name comes from the English terms jump and conditional. The jump is executed if the current result is '1'.

The jump destination is specified by a mark or label. The label is inserted before the command at which processing is to be continued in the case of a jump. You can freely choose the name of the label. Please note: the colon is *only at the destination point* after the label.

The sequence of this decision can be represented graphically in different ways: either with the *flowchart* as in Fig. 10.1 or with the *structure chart* (Fig. 10.2). In the flowchart, the jumps can be easily traced, whereas the structure chart allows a more compact representation. For the "instruction list" language used in this book, the program representation in the flowchart is usually better suited than the structure chart.

`jmpc Mark` Check and jump to label if necessary (*without* colon)
`Mark:` Definition of jump target (*with* colon)

10.3 Set Initial Value

And another problem:

At the start of the program all values are set to 0 and thus all lights are off. Multiplying is of no use then, because zero times two always remains zero: all lamps remain off; no light point would run. How do you get the initial value 1, i.e. the value with which the multiplication should start?

At the beginning the first light point is to be set. The instruction s `%qx1.0` must be executed "at the beginning" to set the lowest bit in the byte `%qb1`. But it is wrong to execute this in *every* cycle, rather this instruction must be executed only once *directly after the program start*.

This can be achieved by the following part of the program:

```
ldn Cycle1
s   Bit0
s   Cycle1
```

The flag `Cycle1` is '0' at the start of the program. Therefore, the following two instructions are executed: both flag `Cycle1` and bit `Bit0` are set. The `Cycle1` flag is already set in the second cycle, the two following instructions are no longer executed and bit `Bit0` is not set.

With this procedure you achieve the conditional execution of set instructions with the condition: very first cycle run.

10.4   Running Light

Exercise 10.1 (FLASH101)

Create a chaser program. Use the suggested variables:

```
var
  Pulse: TON
  Cycle1: BOOL
  Bit0  AT %QX1.0: BOOL
  Bit7  AT %QX1.7: BOOL
  Byte1 AT %QB1: USINT
end_var
```

In the suggested sample solution, make sure that you set the "Standard I/O" process to byte 1.

In this solution, note the allocation of a BOOL type to a portion of the memory of the USINT number `Byte1`.

10.5   Comparisons

Restart in the Same Running Direction

When the running light from Exercise 10.1 has run through, the program terminates and you get a runtime error: when the last LED is lit is the content of the USINT variable. $1000\ 0000_{bin}=128_{dez}$. This multiplied by 2 gives 256, which is outside the value range of USINT.

If the running light is to start again and again from the beginning, the top bit must be deleted in the case %QB1 = 128 and the bottom bit must be set again. You can test whether the current result has the value 128 with the EQ operator.

```
LD Byte1
EQ 128
R  Bit7
S  Bit0
```

To effectively prevent the overflow, this code must be executed *before* the mul   2 instruction.

EQ number tests whether the current result is equal to the number.

However, you will now notice that the running light does not start with the first but with the second LED when restarting. Of course: after setting bit0, mul 2 is executed immediately, so the value is immediately set to 2_{dez}.

How can you correct this now? – One possibility is to *reset* the flag Cycle1 instead of setting Bit0. This will restart the program exactly as it did immediately after power-up. The instruction mul 2 will still be executed when restarting, but then it has no effect any more, because after deleting Bit7 the numerical value in Byte1 is 0.

Exercise 10.2 (FLASH102)

Make sure that the running light starts again from the beginning when the last LED lights up!

Change of Direction

It can now also be desired that the running light changes its direction when reaching the end and runs back again. We will do this in the next section, but here is a preliminary exercise: The running light should run in the other direction, starting with the top bit %qx1.7. Shifting the bit to the right *halves* the value of the byte, so you have to divide the output value by two.

The div <number > operator divides the current result by the number.

Exercise 10.3 (FLASH103)

Create a right-hand running light!

Fig. 10.3 Structure diagram
for clockwise/anticlockwise
rotation

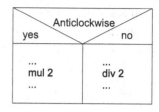

10.6   Unconditional Jump

It is possible that a certain part of the program must be skipped in *any case*. You can't
imagine that? You think, then you could omit this part right away? In the following we will
show you exactly this case with an example!

Example 10.2 (Right-Left Running Light)

A running light can run both to the right and to the left. In one case you have to divide
by 2, in the other case multiply by 2.   ◀

In the graphical representation (Fig. 10.3), there are two parallel paths. In one of them
you have to multiply, in the other one you have to divide. This looks very simple in the
graphic of the structure chart, but is a real problem in the "Instruction list" language:
Unfortunately, it is not possible to create *parallel branches* in the instruction list, but
everything must always be programmed nicely one after the other. You must implement
such a "parallel structure" in the statement list by means of jumps.

If the running light is currently running left, the program does *not* execute the *conditional* jump to the label `right:` but continues working from the label `left:` and executes the multiplication part (Fig. 10.4). Subsequently, the division part (right run) *must* be
skipped. Exactly here an *unconditional jump* to the label `OK:` is required.

In the statement list language, the `jmp <mark>` operator performs an unconditional jump.

Exercise 10.4 (FLASH104)

Create a back and forth running light! The running direction should be adjustable with
the switch at input `%ix1.0`. Alternatively, you can also try to switch the direction at
the edges in each case.

The following is an excerpt from the program. We have only given the commands
for the decision. You should program the commands for switching between clockwise
and counterclockwise rotation yourself.

Fig. 10.4 Flow diagram for right-left run: The "lower" part must always be skipped if the "upper" part has been carried out!

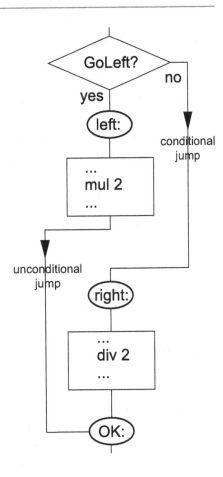

```
    ldn GoLeft
    jmpc Right (* conditional jump *)
Left:          (* Left rotation    *)
    ...
    jmp OK     (* unconditional jump*)
Right:         (* Right rotation    *)
    ...
OK:
```

Exercise 10.5

The PLC provides various "rotating commands" with which the realization of running lights is also possible. Table 16.9 lists the functions included in PLC-lite. Try to program running lights with SHL (Shift left by N bit, fill right with zero), SHR (Shift right by N bit, fill left with zero), ROL (Rotate left by N bit, "in a circle") or ROR (Rotate right by N bit, "in a circle")!

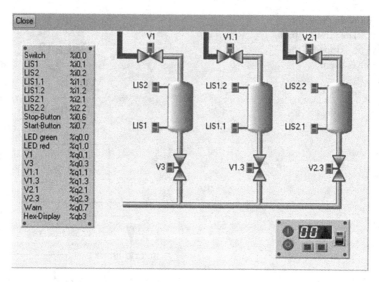

Fig. 10.5 Large "Tanks (big)" measurement model

10.7   Filling Several Measuring Vessels

Figure 10.5 shows the "Large measurement model". The three boilers are to be filled one after the other. The valves V1, V1.1 and V2.1 must be actuated one after the other for this purpose. This is practically the same as a running light. In contrast to the tasks discussed earlier with the running light, however, it is now not a time cycle that determines the switching on, but a signal from the process: Reaching the upper filling levels LIS2, LIS1.2 or LIS2.2 in each case gives the start signal for filling the next boiler.

Exercise 10.6 (TANK102)

After the start signal with the 'I' key, the three boilers are to be filled and emptied again individually one after the other. Use the function block FB_Tank.IL for the tank fillings. Test the program with the large measuring model "Tanks (big)".

10.8   Multiple Selection

Example 10.3 (Control of a 7-Segment Display)

In a 7-segment display, seven LEDs are arranged so that the combination of several light bars represents a digit in readable form. (see Fig. 10.6) The LEDs light up when a '1' is present at the relevant terminal of the module. The block has 8 connections: seven

Fig. 10.6 Designations of the segments of a 7-segment display

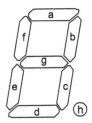

for the segments and one for the dot. The segments are labeled clockwise. In the "7-segment" process model from PLC-lite, segments a to h are connected to outputs %q0.0 to %q0.7, i.e. a to %q0.0, b to %q0.1, and so on. ◄

Figure 10.7 shows the displays of the digits 0 to 9 and the letters A to F. You can use the assignments in Fig. 10.6 to determine the segments to be controlled for each digit.

Exercise 10.7 (7SEG101)

Create a table that shows which segments have to be controlled for the different characters.
  Check the table with PLC-lite! (Process: "7-Segment").

Exercise 10.8 (7SEG102)

By means of the input switches at %ib1, digits are set which are displayed on the 7-segment display. Complete the following part-program! The flow chart is partially given in Fig. 10.8 and the structure diagram in Fig. 10.9.

```
Program SevenSegment
Var
   InByte AT %IB1: Byte
   OutByte AT %QB0: Byte
end_var
   ld InByte
```

Fig. 10.7 Display of HEX digits on a 7-segment display

0?
1?
A?
Zero:
Load Pattern0
One:
Load Pattern1
A:
Load PatternA
End:

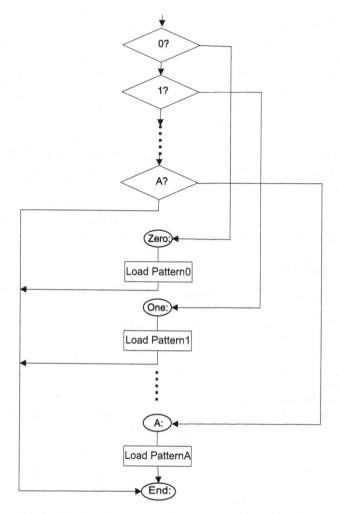

Fig. 10.8 Flow chart for Exercise 10.8

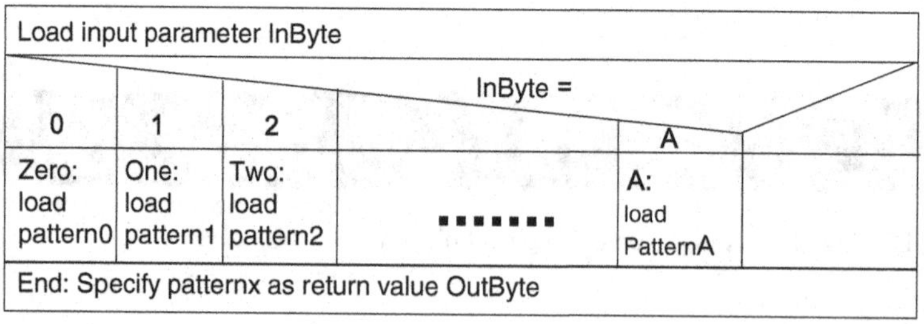

Load input parameter InByte
InByte =
0
1
2
A
Zero: load pattern0
One: load pattern1
Two: load pattern2
A: load PatternA
End: Specify patternx as return value OutByte

Fig. 10.9 Structure chart for Exercise 10.8

```
   eq 0
   jmpc Zero
   ld InByte
   eq 1
   jmpc One
   ld InByte
   eq 2
   jmpc Two
...
   jmp End
Zero:
   ld 2#001111
   jmp End
One:
   ld 2#00000110
   jmp End
Two:
...
End:
   st OutByte
end_program
```

10.9 Random Numbers

Here we present a function block that outputs "random numbers" between 1 and 6. The pulses of a fast running generator are added up in a counter. When 5 (= sixth number, because the counter counts from zero) is exceeded, the counter is reset. The pulses are counted as long as the input value run is '1'. This input value can be read by a push button. If the cycle frequency of the PLC is sufficiently high, the generator works so fast that the counter runs several times from 0 to 5 when a button is pressed (however briefly). The counter value is thus quasi "random".

In the following Example 10.4 the generator is programmed tricky: to get the fastest possible frequency, no timer is used, but the boolean variable toggle is "flipped" in every cycle by first loading it negated with ldn toggle and immediately storing it again with st toggle. This value is written to the counter input. Thus the counter receives a clock pulse with two cycles.

When the counter has reached the count value 6, it must be reset, because the counter should count impulses from 0...5 = 6. After setting the R-input the counter must be called again with cal Counter, so that it is set to zero immediately.

Example 10.4 (Function Block for Generating Random Numbers)

```
Function_Block Chance16
var_input
  run: BOOL
end_var
var_output
  Value: SINT
end_var
var
  Counter: CTU
  toggle: BOOL
end_var
  ldn run
  jmpc output
(* Generate counting pulses *)
  ldn toggle
  st toggle
  st Counter.CU
  cal Counter
(* Check final value *)
  ld Counter.CV
  gt 5
  st Counter.R
  cal Counter
issue:
  ld Counter.CV
  int_to_sint
  add 1
  st Value
end_function_block                                                    ◄
```

Exercise 10.9 (DICE101)

Draw a flow chart for the function block Chance16. Write down the function block, save it as file FB_CHANC.IL. Create a main program which outputs the random numbers on the HEX display at %qb1 after pressing the key at %i0.0.

Note: In PLC-lite you can set the cycle time of the PLC in the configuration window. Later, in Sect. 11.3, we will also display the numbers in "dice form"!

Functions

11

Abstract

You are probably familiar with functions from mathematics. A function is 'passed' a number, which it processes and 'returns' the result (again, a number!). For example, the function 'square' returns the square of every number entered. In this chapter you will learn how functions are applied in the PLC according to IEC 61131-3.

We will show you how to use functions in IEC 61131-3 using the example of a converter for digits to the 7-segment display (seven-segment decoder).

11.1 Use of Functions

A function returns an output value for an input value. The input value can be a number, e.g. a byte. The output value can be a different representation of this number, for example in HEX code or as a control code for a numeric display (7-segment code). It is advantageous to use the digit to 7-segment display converter from Sect. 10.3 as a function.

We will explain the use of a function using the example of converting a byte value into a 7-segment code.

The function is given the name `ByteTo7Seg`, the associated main program is to be called `SevenSegment`. In the main program the function is called with its name (`ByteTo7Seg`).

The calling main program reads the input value into the Current Result CR with the `ld` operator and then calls the function. It receives the return value of the function again in the Current Result CR. It writes this current result (which has changed in the meantime) into the output byte `%qb0` using the `st` operator.

© The Author(s), under exclusive license to Springer-Verlag GmbH, DE, part of
Springer Nature 2022
H.-J. Adam, M. Adam, *PLC Programming in Instruction list according to IEC 61131-3*, https://doi.org/10.1007/978-3-662-65254-1_11

Example 11.1 (Main Program with Function Call)

```
Program SevenSegment
var
 InByte AT %IB1: Byte
 OutByte AT %QB0: Byte
end_var
 ld InByte
 ByteTo7Seg (* call of the function *)
 st OutByte
end_program
```
◄

Now how do you have to create the function itself? Have another look at Exercise 10.8. Compare the program given there with the function IntTo7Seg printed in the following Example 11.2!

Example 11.2 (Function for Controlling the 7-Segment Display)

```
Function ByteTo7Seg: Byte (* type return value *)
var_input
 Digit: Byte (* Declaration input value *)
end_var
 ld Digit (* input value *)
eq 0
 jmpc Zero
 ld Digit
eq 1
 jmpc One
 ...
Zero:
 ld 2#001111
 jmp End
One:
 ld 2#00000110
 jmp End
 ...
End:
 st ByteTo7Seg (* return value *)
ret (* end *)
end_function
```
◄

Can you see in Example 11.2 how the function ByteTo7Seg takes over a numerical value on the variable Digit? This is transferred to the current result with the ld Digit and processed further. Afterwards, the result is allocated to the memory for the return value, which is addressed under the function name (st ByteTo7Seg). The *return* is indicated by RET. In the line Function. . . the data type of the return value is specified after the colon.

Exercise 11.1

Describe in your own words how a function is structured, how it reads the input value from the main program and how it returns the result to the main program!

With the variable between var_input and end_var the function receives its input value. The RET instruction marks the end of the function.

The function must be placed in its own file (like a function block). You must also have a "main program" block in your project that calls the function.

Exercise 11.2 (7SEG111)

Create the main program and the function for the 7-segment coding as in Example 11.2! The program is to convert the input byte of %ib1 into a seven-segment form. The 7-segment display is to be connected to %qb0.

11.2 Difference Between Function and Function Block

Unfortunately, in IEC 61131 the names "function" and "function block" are easy to confuse. Therefore, pay close attention to the difference:

The Function Block contains a "*memory*". The output of a function block can take on different values despite the same input signals. This may sound a bit puzzling, as if random values were output by a function block. No, the output values of a function block are not random! Let's consider a counter. The pulse to be counted always produces the same input signal; but depending on the previous process, the function block outputs a different numerical value. This means that the function block must have a memory in which it stores the previous numerical value in order to be able to determine the next number after a counting pulse.

For the Function the output value *always* depends *directly* on the input value. In Example 11.2, the same output value is always output for the same specific input value, regardless of which value the function processed in a previous call. The function does not need a memory, because the output value can always be determined from the current input value.

Unlike function blocks, functions do **not** have to be instantiated separately before use. The PLC operating system automatically occupies the memory required for the variables of the function only while it is called. After returning from the function, this memory area is immediately released again. This means that all variable contents changed within the function are no longer available the next time the function is called, and this results in the typical behavior described above for a function: it "forgets" old values.

> The function block contains a "*memory*".
> The output value of a *function* always depends directly on the input value.

11.3   Dice Game

In Exercise 10.9 you programmed a dice game with a random number generator. Now we want to extend it with a nice dice display. The function for converting the number values into the display works practically the same as converting a number into the 7-segment code!

Exercise 11.3 (DICE111)

Extend Exercise 10.9 by adding the `IntToDice` function to display the dice results in the Dice process. Save the function under the name `F_Dice.IL`.

11.4   BCD Converter

In Exercise 8.15 the boiler content should be displayed in litres. At that time we had the "blemish" that the display was in hexadecimal numbers. We now want to change this and design a function that converts the hexadecimal notation into a BCD number notation.

Exercise 11.4 (BCD111)

Create a function to convert hexadecimal numbers into BCD numbers! Test the function in a suitable program, for example the one from Exercise 8.15.

11.5   Parameter Transfer to the Function

The functions used so far formed an output value from exactly *one* input value. The input value was brought into the function via the current result and the output value was returned to the calling program via the current result.

How does this work if the function requires more than a single input value?

For example, the function is to determine the larger of two numbers and return it. In the PLC it works like this: The first number is loaded into the Current Result CR with the LD operator, the second number is written as a parameter after the function name.

Perhaps you notice the similarity of a function to an operator (e.g. add)? In both cases, one operand is loaded into the Current Result, while the second is passed after the function name. The calculation result is then available in the Current result!

> A function returns exactly one data element after the call. In the instruction list language, the return value is in the Current Result (CR). It differs little in use from an operator.

Example 11.3 (Function Call with Passing of Two Parameters)

```
MaxTest program
  ...
  ld Digit1 (* Load parameter1 in CR *)
  Greatest Digit2 (* call and load param. 2 *)
  st OutDigit (* Accept return value *)
end_program                                              ◄
```

The function itself now has two input parameters in the list between var_input and end_var.

- The value listed first is taken from the Current Result (CR),
- the second is the parameter after the function name.

Example 11.4 (Variable Declaration with Two Input Parameters)

```
Function Greatest: int
var_input
  D1: int (* Parameter1 from CR *)
  D2: int (* load parameter2 *)
```

```
 end_var
  ...
 ret (* return value in CR *)
end_function                                                    ◄
```

Exercise 11.5 (MAX111)

Write the function `Greatest`, which returns the larger of the two numerical values passed in, and test it with a suitable program example.

Exercise 11.6 (TIME111)

In response to a start signal, two stopwatches are to run which can be stopped independently of each other using separate buttons. The respective `time1` or `time2` is to be displayed when the button is pressed. The shorter time is to be displayed in `byte2`, the longer time in `byte3`.

If more than two values have to be passed, the parameter list can be extended after the function name. Separate the individual parameter values with commas. For example, a function is to check whether a number lies between two limit values. This so-called discriminator function is passed the number to be tested and the two limit values, i.e. a total of three values.

Exercise 11.7 (DISKR111)

Write a function `Between`, which you save as `F_Diskr.IL`. It is to compare two values passed as parameters with the value standing in the current result. The function should return 'true' if the current result is between the two parameter values.

Example 11.5 (Main Program with the Function Call and the Parameter List)

```
program discriminator
var
 CompValue AT %ib0: sint
 MinValue AT %ib1: sint
 MaxValue AT %ib2: sint
 OKLamp AT %qx0.0: bool
end_var
  ...
 ld Compvalue
 Between MinValue, MaxValue
 st OKLamp
  ...
end_program                                                     ◄
```

Sequence Controls

12

Abstract

In the preceding chapters you have learned the language and structural elements for programming the PLC using instruction lists. This gives you the tool to control any type of technical process with PLC.

A certain type of process can be subdivided into a sequence of individual steps that always run one after the other. For these, the programming logic of "sequential controls" is suitable, which in these cases results in a clear program structure. The IEC 61131-3 standard defines a special language for programming sequential controls: the "sequential function chart" (SFC), but this is not the subject of this course. We apply the programming logic behind sequential programs in the Instruction List language. Although this is not really practical, it has the advantage that the basic procedure can be easily understood.

12.1 Basic Principle of Sequence Control Using the Example of Pushbuttons

The sequence controls are so important in practice that the standard provides a special language for them. In this chapter, however, you will not implement the sequence control with this special language, but in the "instruction list" language. This is somewhat more "cumbersome", but you will gain an insight into the working method and the basic ideas behind this programming language.

Most controls can be broken down into successive individual steps. This breakdown provides a better overview. The transition from one single step to the next depends on conditions; this can be a reached filling level or a certain temperature, but also the expiry

© The Author(s), under exclusive license to Springer-Verlag GmbH, DE, part of Springer Nature 2022

H.-J. Adam, M. Adam, *PLC Programming in Instruction list according to IEC 61131-3*, https://doi.org/10.1007/978-3-662-65254-1_12

of a time period. The programs gain considerably in clarity through the realization as sequence control and can therefore be programmed and later maintained much more easily.

You can also imagine the principle of sequence control like this: The process sequence is shown on a display panel. Each important process step is described on a backlit field, e.g.: (1) filling, (2) stirring, (3) heating and stirring, (4) emptying. Then everyone can see the instantaneous working status of the plant at a glance. You see: Only one step is active at a time, i.e. only one field is illuminated. In each step certain actions are executed. The sequence of the steps is predefined and certain conditions must be fulfilled for the change to the next step. The sequence control by means of PLC supports these properties of the process.

Before we bombard you with a lot of theory, let's first consider a simple example to explore the principle of sequential control. The following problem, which at first seems quite simple, is very suitable:

Example 12.1 (Push Button)

A lamp should light up as long as a button is pressed.

This task can of course be accomplished with the simple program:

```
Program PushButton0
VAR
    Button AT %IX0.0: Bool
    Lamp AT %QX0.0: Bool
END_VAR
    LD Button
    ST Lamp
end_program                                                          ◄
```

But we don't want to make it quite that easy for ourselves, after all there are much more complex tasks to master! That is why we will proceed systematically step by step:

The entire process can be broken down into two *steps*, which are shown graphically in Fig. 12.1:

- **Step 1:** The button is not pressed
  - The lamp is off
  - Waiting for keystroke

- **Step 2:** The button is pressed
  - The lamp is on
  - Wait until key is released

If you look at these steps more closely, you can identify an *action* at each step (the lamp is off/on) and a *transition condition* that must be met for the system to move to the next step (wait for key press/wait for release).

Fig. 12.1 Sequence of the
Example 12.1 push button

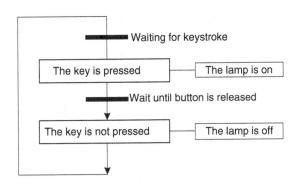

12.2   The Sequence Chain

Example 12.1.(pushbutton) is now transferred to the PLC language:

The control is divided into *steps* and forms a chain of steps. In this *sequence,* only *one* step is active at any time. In order to reach the next step, a *condition (transition)* must be fulfilled. The controller executes different *actions for* each step.

This looks just like a running light! Instead of an illuminated light, we have an active process step. However, the transition from one state to the next is not given by a time clock, but rather as in the problem already discussed in Sect. 10.7: One action after the other runs and the transitions (transition conditions) are provided by the process.

Example 12.2 (Pushbutton as Sequence Chain)

The control for Example 12.1 can be realized with the following program part: To program the sequence chain, we can provide a flag (*"step flag"* Step1 and Step2) for each step. A set flag then means: Step active. The transition is made by resetting the corresponding flag as soon as the *transition conditions* are fulfilled.

```
Program PushButton1
VAR
   Step1: bool
   Step2: bool
   Button AT %IX0.0: Bool
   Lamp AT %QX0.0: Bool
END_VAR
(* Sequence of events: *)
(* STEP 1 *)
   LD Step1 (* Step 1 active? *)
```

```
  AND Button (* button pressed? *)
  R Step1 (* Exit step 1 *)
  S Step2 (* Activate step 2 *)
(* STEP 2 *)
  LD Step2 (* Step 2 active? *)
  ANDN Button (* button released? *)
  R Step2 (* Exit step 2 *)
  S Step1 (* Activate step 1 *)
(* ACTION in step 2 *)
  LD Step2 (* Step 2 active? *)
  ST Lamp (* Switch on lamp *)
end_program                                                  ◄
```

Note that in program Example 12.2, a separation has been intentionally made between the flow of the sequence chain and the action to be executed in step 2. In the example, an action is only required in step 2, but in general one or more actions will have to be executed in each step. The actions to be executed in each step are only programmed at the end, after the sequence chain. This increases the clarity of the program.

12.3 Set Initial State

Take another critical look at program Example 12.2! Concentrate completely on the beginning of the program, at the very beginning, when the program is started. Answer the question: What are the values of the two flags? Exactly: Both are '0'; the control cannot start at all! The control can only start if you make sure that *exactly one* flag is set at the beginning. You have already learned how to do this in Sect. 10.3. We still have to make sure that exactly one flag is set at the start of the program.

```
ldn Cycle1
s Cycle1
s Step1
```

In sequence control, *exactly one flag* must always be set in the sequence chain.

Exercise 12.1 (PUSH121)

Write down the sample program `PushButton1` and test it. The test is positive when the lamp is lit as long as the button is pressed at %IX0.0 and goes out when the button is released.

We can understand very well if you are not yet convinced of the sense of the sequence controls, the program became quite considerably longer than the program Example 12.1 `PushButton0` with the same functionality! But just a little more patience, and you will clearly see the advantages. If you have tried Exercises 6.16 and 7.12, you will soon appreciate the advantages of the new design method!

12.4   Sequence Step and Step-On Condition

A sequence control has a positively controlled sequence. The smallest functional unit of such a control is a *sequence step* (or *step* for short). The entire task is therefore broken down into individual "processing blocks" that can be separated from one another. The sequence of the steps results in the entire program, the so-called *sequence chain*.

A *process element* is assigned to each step. In the example, this is a flag that generates the signals required to influence the process, i.e. the *actions,* via links.

If the step-on *condition* is fulfilled, the next step is activated by setting the next flag and resetting the current one. The step-on conditions can come from outside (keystroke by the operating personnel), from the process itself ("from inside": e.g. temperature reached, fill level reached, time elapsed, etc.) or also from a timer.

Because a flag is assigned to each step, the *current state of the control* can be seen at any time from the flag set.

Care must be taken when setting up the step-on conditions! If, for example, the condition *cannot* occur in the process, then the controller *can* no longer continue to run, it "blocks". A frequent cause for the blocking of a control is also *forgotten* step-on conditions. However, compared to a "normal" control, errors can usually be detected quickly in a sequence control.

12.5   Graphical Representation of Sequence Controls

In practice, sequence programming offers several advantages:

- Simple project planning and programming
- clear program structure

Fig. 12.2 Graphical representation of the sequence chain

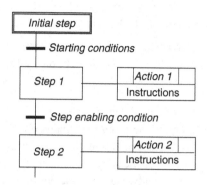

- easy change of the function sequence
- in the event of faults, easy recognition of the cause of the fault
- adjustable different operating modes.

Practically all technical processes can be broken down into sub-steps that must be executed in a specific sequence one after the other. This is referred to as a "sequence". Each
step in this sequence can be precisely designated (e.g. numbered). The steps are divided in
such a way that only one specific step is ever being processed, i.e. is "active".

Actually, the process control for technical processes is not that unusual: When you
describe a technical process, you "automatically" break down the entire process into a
sequence of individual processes.

The IEC 61131-3 standard also describes the sequential function chart (SFC) and its
graphical representation. It is based on the DIN EN 60848 standard, which regulates the
representation of sequential controls. This is known by the acronym "GRAFCET"
(GRAphe Fonctionnel de Commande Etapes/Transitions) and has replaced DIN 40719
Part 6 since 2005. An example is shown in Fig. 12.2.

- The step symbols are displayed as rectangles. The step is described either by a number
  or a symbolic name.
- The commands to be executed when this step is active are displayed as *action blocks* in
  a separate field to the right of the step symbol.
- At the top of the action block, the *action name* is entered in the middle.
- In the lower part of the action block, the commands to be executed with this step are
  specified.
- The entries in the left part above are the *action destination marks (designator)*. We will
  discuss them a little further on.
- In the upper right, optional part of the action block, a "display" variable of data type
  BOOL can be specified to indicate the completion, timeout or error conditions, etc. of
  this action block.

12.6   **Pressure Switch**

The following task already shows the advantage of the systematic approach to programming in the sequential function chart. Perhaps you have dealt with Exercise 6.16?

With the so-called "bedside lamp switch", the light is supposed to change every time the switch (actually: pushbutton) is pressed: The first press switches the light on, which is switched off again with the next press on this very pushbutton. This can no longer be done with an "ordinary" pushbutton! And without a systematic approach, this task becomes a fiddly job!

We can distinguish four different states (steps):

1. Button released, light off, wait for pressure
2. Button pressed, light on, wait for release
3. Button released, light also on, waiting for pressure
4. Button pressed again, light off, wait for release

The light must be on in both step 2 and step 3. So in these two steps the action "light on" has to be executed, in the other two nothing.

Exercise 12.2 (PUSH122)

A pushbutton is to switch on a lamp; by pressing this pushbutton again, the lamp is to be switched off again.

Fig. 12.3 Sequence chain to the pressure switch (Exercise 12.2)

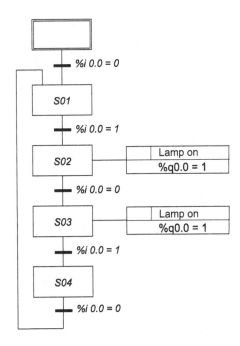

The control for this task can be realized with the program Example 12.3 and the corresponding sequence Fig. 12.3.

```
Program PushButton2
VAR
   Step1: bool
   Step2: bool
   Step3: bool
   Step4: bool
   Cycle1: bool
   Button AT %IX0.0: bool
   Lamp AT %QX0.0: bool
END_VAR
   ldn Cycle1
   s Cycle1
   s Step1
(* Sequence of events: *)
(* STEP 1 *)
   LD Step1 (* Step 1 active? *)
   AND button (* button pressed? *)
   R Step1 (* Exit step 1 *)
   S Step2 (* Activate step 2 *)
(* STEP 2 *)
   LD Step2 (* Step 2 active? *)
   ANDN button (* button released? *)
   R Step2 (* Exit step 2 *)
   S Step3 (* Activate step 3 *)
(* STEP 3 *)
   LD Step3 (* Step 3 active? *)
   AND button (* button pressed? *)
   R Step3 (* Exit step 3 *)
   S Step4 (* Activate step 4 *)
(* STEP 4 *)
   LD Step4 (* Step 4 active? *)
   ANDN button (* button released? *)
   R Step4 (* Exit step 4 *)
   S Step1 (* Activate step 1 *)
(* ACTIONS *)
   LD Step2 (* Step 2 active? *)
   ST Lamp (* Switch on lamp *)
   LD Step3 (* Step 3 active? *)
```

```
   ST Lamp (* Switch on lamp *)
 end_program
```
◀

12.7 Control of Actions

With the momentary-action switch, the lamp must be activated in steps two *and* three, i.e. the same action must be performed in two steps. In the above example, the two flags for the respective steps are linked with OR.

```
alternatively for the ACTIONS:
(* ACTION *)
  LD Step2 (* Step 2 active? *)
  OR Step3 (* Step 3 active? *)
  ST Lamp (* Switch on lamp *)
```

The same objective can be achieved by *setting* the lamp to "ON" in step two and *resetting* the lamp to "OFF" in step four. The action thus extends over several steps. In such cases, the PLC must remember the action over several steps, i.e. save it. For saving, the output can be set or reset directly. This is shown graphically in the sequence in Fig. 12.4.

> If an action is to be effective over several steps, it must be executed "saving".

In the step in which the action begins, the output is set; in the step in which the action is to end, the output is reset.

Some commands should only be active during the current step ("non-storing"), others must have an effect over several steps ("storing"), still others must last for a very specific time before the step counter is allowed to continue.

When issuing commands, the non-storing (N) commands are the easiest to handle. The execution of the command stops as soon as the step is exited.

Table 12.1 lists some of the actions defined in the IEC 61131 standard. The type of action is indicated by the action designation mark (designator) in the left part of the command field.

Exercise 12.3 (PUSH123)

Change the program for the pushbutton switch (Example 12.3) so that in step 2 the lamp is switched ON and in step 4 the lamp is switched OFF (Fig. 12.4). You can extend the program so that the respective step flag is displayed on LEDs 1.1 to 1.4.

Fig. 12.4 Sequence of events
for Exercise 12.3 with setting/
resetting (storing) of the states

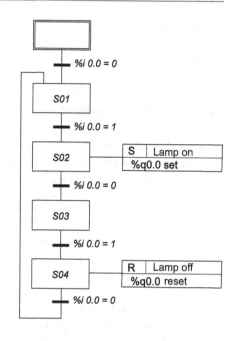

Exercise 12.4 (TANK121)

Create the sequence control for the system in Fig. 12.5 according to the flow diagram!
Enter the instruction list printed in Example 12.4 into the PLC, test the program and
describe verbally (in words) the process sequence.

Example 12.4 (Tank)

Proposed programme for Exercise 12.4:

```
Proposed programme for Exercise 12.4:
Program Tank
VAR
   S01: BOOL
   S02: BOOL
```

Table 12.1 Action designator

Designator	Action, explanation
None or N	Non-storing
R	Overriding reset
S	Set stored
L	Time limited
D	Time delayed

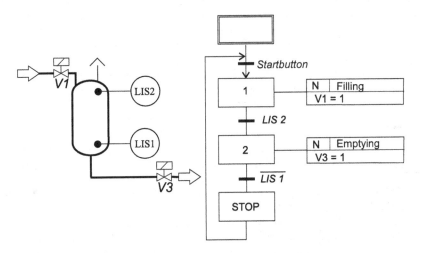

Fig. 12.5 System and flow diagram (sequencechart) for Exercise 12.4

```
  S03: BOOL
  Cycle1: BOOL
  V1 AT %qx0.1: BOOL
  V3 AT %qx0.3: BOOL
  LIS1 AT %ix0.1: BOOL
  LIS2 AT %ix0.2: BOOL
  Start AT %ix0.7: BOOL
END_VAR
  LDN Cycle1
  S Cycle1
  S S03
(* STEP 1 *) (*Drain chain *)
  LD S01 (*step 1 active? *)
  AND LIS2 (*full? *)
  R S01 (*exit step 1 *)
  S S02 (*activate step 2*)
(* STEP 2 *)
  LD S02 (*step 2 active? *)
  ANDN LIS1 (*empty? *)
  R S02 (*exit step 2 *)
  S S03 (*activate step 3*)
(* STOP *)
  LD S03 (*stop step? *)
  AND Start (*Start button? *)
  R S03 (*exit step 3 *)
```

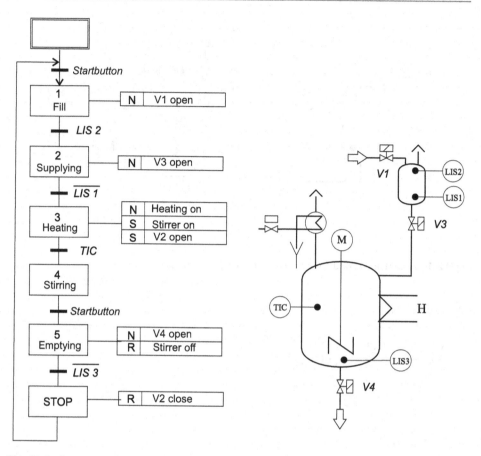

Fig. 12.6 Sequence and apparatus for Exercise 12.6

```
  S S01 (*activate step 1*)
(* ACTION.)
  LD S01 (*step 1 active? *)
  ST V1 (*open V1 *)
  LD S02 (*step 2 active? *)
  ST V3 (*open V3 *)
end_program                                    ◄
```

Exercise 12.5 (TANK122)

Improve the program from solution 12.4 so that the boiler is only filled and emptied once, even though the start button is pressed continuously. Before a new filling process, the start button must first be released!

Fig. 12.7 For Exercise 12.7:
Changing the sequence by time
elements

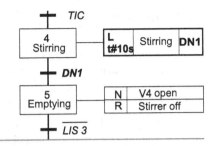

The plant set-up shown in Fig. 12.6 is to be operated with a sequence control. Compare
the functional diagram with the verbal description. Implement the control system with
the aid of PLCs!

Verbal Description:

1. Water is supplied to a measuring vessel via valve V1 until the limit switch LIS2
   signals "measuring vessel full".
2. Via valve V3, the water supplied is drained into the stirring piston until LIS1
   signals "measuring vessel empty".
3. The flask content is heated up until the contact thermometer TIS indicates "Set
   temperature reached". When heating up, the stirrer must be switched on and the
   cooling water valve V2 opened.
4. Once the set temperature has been reached, the flask contents continue to be stirred.
5. If stirring has been "long enough", the stirrer should be switched off by pressing
   the start button again and valve V4 opened to empty the piston.
6. After emptying the piston (LIS3 = 0), the cooling water is closed and the entire
   process can be repeated by pressing the "Start" button.

12.8   Programming the Timing Elements in a Sequence Control System

In the previous control, in step 4, the post-stirring time should be automatically terminated
after 10 s without the need to press the button again. Figure 12.7 shows the change in the
flow chart for steps four and five.

Complete the control with the timer for the transition from step 4 to step 5! Why does
it make sense to use the timer with switch-on delay (TON)?

Note the ease with which flow control can be extended: see Exercise 13.28 for a task
description for an extended process.

12.9   In Conclusion

Congratulations and our appreciation for your perseverance and achievement! You have now reached the end of this introductory course. We hope you have enjoyed the work and can use the suggestions for your own tasks.

In order to support you in your further programming exercises, we have compiled some repetition tasks in the following chapter. In the last chapter, essential features of "PLC-lite" are compiled, so that you can clarify further questions yourself with the help of this reference.

We wish you continued success!

Repeat Tasks

13

Abstract

This chapter contains further practice and revision exercises with which you can consolidate the knowledge you have acquired.

13.1 Review Tasks for Chap. 2

Exercise 13.1 (LOGIC27)

Give the functional equation for all four switching elements Fig. 13.1!
Complete the function table Fig. 13.2!
   To which of the links drawn above does the given function table belong? (Fig. 13.3).

Exercise 13.2 (LOGIC28)

Which of the switching elements Fig. 13.4 can replace the combination of the two logic elements Fig. 13.3?

Exercise 13.3 (LOGIC29)

A lamp L is to be controlled so that it lights up whenever only switch S_1 or only switch S_2 is operated. Figure 13.5 shows the function table to be implemented.
   Which of the functional elements Figs. 13.6, 13.7, 13.8, or 13.9 satisfies these conditions? What are the functional equations in each case?

Exercise 13.4 (LOGIC30)

What is the functional equation for the circuit Fig. 13.10? Try to make the function table. Since we have 5 variables here, the table must have 32 rows. However, you can

© The Author(s), under exclusive license to Springer-Verlag GmbH, DE, part of
Springer Nature 2022
H.-J. Adam, M. Adam, *PLC Programming in Instruction list according to IEC 61131-3*, https://doi.org/10.1007/978-3-662-65254-1_13

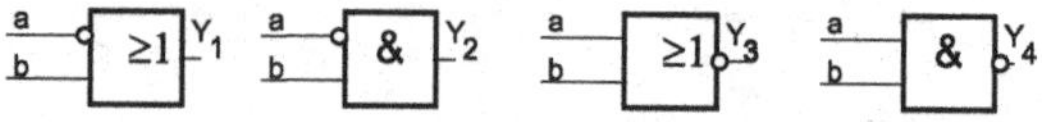

Fig. 13.1 Switching elements
for Exercise 13.1

Fig. 13.2 Function table for
Exercise 13.1

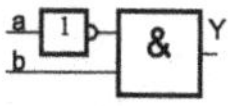

a	b	Y	Y_1	Y_2	Y_3	Y_4
0	0	1				
1	0	0				
0	1	0				
1	1	0				

Fig. 13.3 Switching element for
Exercise 13.2

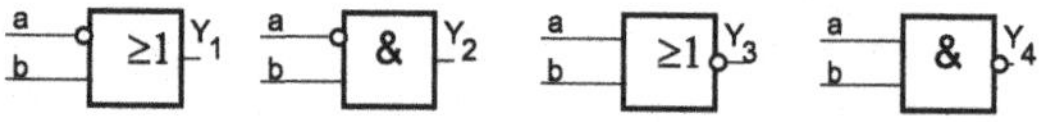

Fig. 13.4 Switching elements
for Exercise 13.2

Fig. 13.5 Function table for
Exercise 13.3

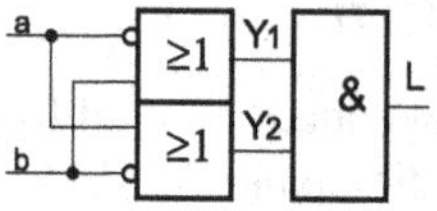

S_1	S_2	L
0	0	
1	0	
0	1	
1	1	

Fig. 13.6 Function 1 for
Exercise 13.3

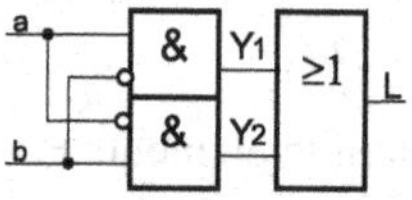

a	b	y_1	y_2	L
0	0			
1	0			
0	1			
1	1			

Fig. 13.7 Function 2 for
Exercise 13.3

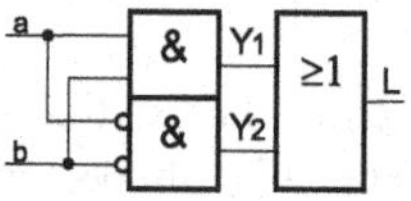

a	b	y_1	y_2	L
0	0			
1	0			
0	1			
1	1			

Fig. 13.8 Function 3 for
Exercise 13.3

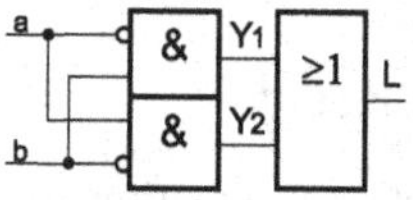

a	b	y_1	y_2	L
0	0			
1	0			
0	1			
1	1			

Fig. 13.9 Function 4 for
Exercise 13.3

a	b	y_1	y_2	L
0	0			
1	0			
0	1			
1	1			

Fig. 13.10 Circuit for
Exercise 13.4

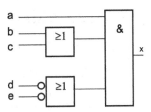

reduce the number of rows to be written if you "trick" a little: for $a = 0$, namely, x always $= 0$!

Exercise 13.5 (FLASH23)

A lamp should be able to be switched either to off, continuous light or flashing light. Provide one switch as on/off switch and a second switch for switching between continuous and flashing light. To solve this task, you should refer to Sects. 2.7: AND operation as data switch and 2.9: OR operation as data switch.

13.2 Review Tasks for Chap. 5

Exercise 13.6 (TEMP51)

The temperature of a chemical process is monitored with a bimetallic thermometer. If the temperature falls below a certain value, the signal transmitter reports this with the signal value '0' and an alarm horn is actuated.

Exercise 13.7 (LOGIC51)

In an injection moulding machine, the plunger only moves down when the mould pressure is built up, the protective grid is down and the pressing temperature has been reached. The logical behavior of the individual sensors:

```
Form closed:                     log. 1 (proximity switch)
Forming pressure reached:        log. 0 (strain gauge)
Protection grid bottom:          log. 1 (limit switch)
Pressing temperature reached:    log. 0 (thermocouple)
Punch extends:                   log. 1 (valve)
```

Exercise 13.8 (LOGIC52)

The water supply to a turbine is blocked if a certain speed is exceeded or the bearing temperature is too high or the cooling circuit is no longer in operation. All sensors give log. 1 when the operating condition is no longer fulfilled.

Exercise 13.9 (BOILER53b)

Exercise 5.8 can be extended even further:

(a) The stirrer should run when the switch Sw1 has switched on the system with the heating and the temperature of 50° C has not yet been reached. When the temperature of 50° C is reached, the heating should switch off.

(b) The stirrer should run when the system is switched on with switch Sw1 and the temperature is exceeded 40° C . The stirrer should also continue to run when the temperature reaches 50° C.

Exercise 13.10 (BOILER55)

In a reaction vessel, a safety valve must be opened when the pressure is too high or the temperature is too high or the inlet valve is opened or a certain concentration of the chemical reaction is reached.

```
Pressure too high:      log. 0
Temperature too high:   log. 0
Valve open:             log. 1
Concentration reached:  log. 1
```

13.3   Review Tasks for Chap. 6

Exercise 13.11 (TANK66)

Extension of Exercise 6.15:
   The "Fill, Empty" cycle should be repeated after pressing a start button until a second "Stop" button stops the process again at the end of the cycle, i.e. when the container has run empty!

Exercise 13.12 (COOLER61)

One unit is cooled by two fans. The function is monitored by one air flow monitor each. If both fans fail while the unit is switched on, an acoustic signal should be emitted.
   This message is to be output until the fault message is acknowledged via an acknowledgement button.
   However, the acknowledgement should only become effective when at least one of the two fans is back in operation or the unit is no longer switched on.
   Determine the assignment tables, the control for the signaling device and implement the control with PLC.

Exercise 13.13 (SORTER61)

Long and short workpieces are transported in any order on a conveyor belt. The belt switch is to be controlled in such a way that the arriving parts are selected according to their length and fed to separate delivery stations. The length of the parts is determined by a scanning device (roller lever valves). If a long part passes through the scanning device, all three roller lever valves are briefly actuated. In the case of a short part, on the other hand, only the middle valve is actuated.

Create assignment tables, the function block diagram, the instruction list and implement and test the circuit with PLC.

13.4 Review of Chap. 7

Exercise 13.14 (FLASH74)

Program a generator to output a pulse train at output %q0.0 that gives a total duration $T_1 + T_2 = 5s$ and a duty cycle $k = 3/5$.

What is the frequency of the generator?

The generator should be able to be switched on by pushbuttons with input %i0.7 and switched off with input %i0.6.

Change the duty cycle to $k = 0.25$ at the same frequency!

13.5 Review Exercises for Chap. 8

Exercise 13.15 (COUNT85)

In a floppy disk factory, 10 floppy disks are to be packed in each carton. After filling a carton, it must be replaced by an empty one, which is then filled again. Design a control system! Create the assignment list and the logic diagram, program and test your solution.

Exercise 13.16 (COUNT85b)

Modify the program from Exercise 13.15 so that you can set the number of floppy disks on the digital input unit.

Exercise 13.17 (MIXER82)

The program of Exercise 8.9 is to be extended to control the complete plant set-up shown in Fig. 13.11 consisting of measuring and reaction vessel.

- After filling the measuring vessel twice, the stirrer and the heater should be switched on.
- The stirrer should run for 5 s.
- The heating is switched off as soon as TIC reports the setpoint. At this moment the substance is also drained via V4.

Fig. 13.11 Process model

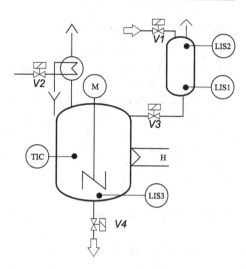

- The process may only be restarted after all the liquid has been drained via V4.

Realize the control with the help of PLC and test the circuit!

13.6 Review of Chap. 9

Exercise 13.18 (FFFB94)

Now use the function blocks FFSR and FFRS (FB_FFSR.IL and FB_FFRS.IL) from both Exercises 9.6 and 9.7 together in one project. Both flip-flops are to be reset and set with the same inputs (%IX0.0 and %IX0.1). Give the outputs to the lamps at QX0.0 and QX0.1. Test the operation of the two flip-flop types in comparison.

Exercise 13.19 (GEN92)

Another example results from Exercise 8.12. There, two generators should be used. Simplify the project by using a function block for the pulses. Name this block 'FB_PULS.IL'.

13.7 Review of Chap. 10

Exercise 13.20 (TANK101)

Do you remember the tasks with the quantity measurement from Sect. 8.9? The CV output of the counter constantly indicates the fill level. By comparing the CV output with a setpoint, a very specific quantity can be filled into the vessel.

Supplement Exercise 8.15 with an input option for the setpoint. As soon as the actual value has reached the setpoint, the filling should stop.

Exercise 13.21 (7SEG103)

In Exercise 10.8 the digits to be displayed were set by means of the input switches at %ib1. Now provide a counter whose count is displayed on the 7-segment display.

Exercise 13.22 (FLASH105)

Let a running light run back and forth over *two* bytes! To do this, load the standard I/O process twice and set the correct byte number for each.

Exercise 13.23 (DICE102)

Play with two dice: Create a program in which two random generators are started and the numbers are output on the outputs %qb0 and %qb1.

Exercise 13.24 (DICE103)

The two random numbers from Exercise 13.23 above are displayed in two different bytes. However, for both bytes only one digit each, i.e. half a byte, namely the "lower" half, is used. Now both half bytes are to be displayed in a single byte.

How can we combine the two random numbers so that both numbers are displayed next to each other in a single byte? Exactly, first the second number must be shifted four places to the left and then the values must be "superimposed".

As already applied to the running light, shifting by one place to the left is done by multiplying by two. But with which number must be multiplied for a shift by four places? In the case of the running light, we multiplied by 2 for a shift of one place. Therefore, for four places, we must multiply by two four times in succession, that is, multiply by $2*2*2*2 = 2^4 = 16$. (Not by $2*4 = 8$).

To finally connect the two values, we only need to OR them. To understand this last step, you should take another look at the "data switch with the OR element" from Sect. 2.9. In the PLC, the OR operator acts on the entire byte and links the bits in the same positions individually.

```
ld Dice0.Value
mul 16
or Dice1.Value
st OutByte1
```

13.8   Review of Chap. 11

Exercise 13.25 (7SEG112)

Modify Exercise 11.2 so that now the upper four input bits are evaluated and appear in
the 7-segment display at %qb1.

In Exercise 11.2, only the lower four bits of the input byte are evaluated. We can also
use the upper four bits. The easiest way to do this is to shift all bits four positions to the
right. Yes exactly, similar to the running light and the cube (Exercise 11.2), but the other
direction: each bit shifted to the right must be divided by two, i.e. by $2*2*2*2 = 2^4 = 16$.

```
ld InByte
div 16
ByteTo7Seg
st OutByte1
```

Exercise 13.26 (7SEG113)

Combine the two seven-segment displays for the low and high parts of the input byte at
%ib1 so that the byte is displayed in two digits.

Now we want to combine the two tasks, and display the input byte of %ib1 in two
digits, each in a seven-segment-digit. So that the two parts: the four low bits and the
four high bits do not interfere with each other, the other half byte must be "hidden", i.e.
the corresponding digits must be replaced with '0'. To do this, you can use the "data
switch with the AND element" from Sect. 2.7.

The PLC applies the AND operation to each bit of a byte individually. The "four-bit"
data switch can therefore be implemented as shown in the following program excerpt.

```
ld InByte
and 2#00001111
ByteTo7Seg
st OutByte0
```

Exercise 13.27 (DICE112)

Create a program for two dice, each using the `ByteToDice` function to display the dice results in the Dice process. Compare here also with Exercises 13.23 and 13.24.

13.9 Review of Chap. 12

Exercise 13.28 (MIXER123)

Extension of flow control Exercise 12.7:

The process of Exercise 12.7 is to be extended. In the following task, only the advancing conditions for steps 3 and 5 must be changed in such a way that they only take place with the execution or expiry of the respective preceding instruction 1. These delay times in turn start depending on LIS1 or LIS3. The necessary changes are marked in **bold in** the following text and in the flow chart Fig. 13.12.

1. Water is supplied to a measuring vessel via valve V1 until the limit switch LIS2 signals "measuring vessel full".
2. Via valve V3 the supplied water is drained into the agitator piston.
   b. If LIS1 signals "measuring vessel empty", valve V3 should remain open for another 2 s so that the measuring vessel can run empty.
3. The flask contents are heated up until the contact thermometer TIC indicates "Set temperature reached". When heating up, the stirrer must be switched on and the cooling water valve V2 opened.
4. After reaching the set temperature, the contents of the flask are stirred for 10 s.
5. When the post-stirring time is finished, the stirrer is switched off and valve V4 is opened to empty the piston.

Fig. 13.12 Extension of an existing sequence control. See Exercise 13.28

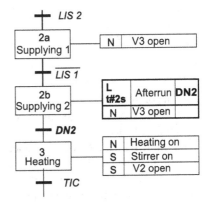

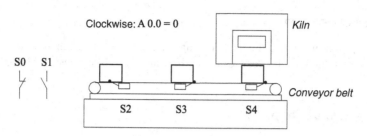

Fig. 13.13 Technology diagram for Exercise 13.30

> **b. After emptying the piston (LIS3 = 0), valve V4 should remain open for another 2 seconds to allow the piston to run empty.**
6. Then the cooling water is closed and the whole process can be repeated by pressing the "Start" button.

- *The transition from step 2 to step 3 (or from step 5 to step 6) only takes place when the signal LIS1 (or LIS3) has gone AND the time of 2 seconds has elapsed.*

Carry out the program extension for steps 2b and 5b! Note that valves V2 and V4 must now be open during several steps.

Exercise 13.29 (MIXER124)

Finally, program a sequence control for the large chemistry process. Let the three measuring vessels fill and empty one after the other. Then the mixture is to be stirred and heated and the waste gas cooled. When the set temperature is reached, the substance is drained and the process can be restarted by pressing the start button. Do not forget the run-on times!

13.10 Mixed Tasks

Exercise 13.30 (EX01)

Kiln control (Fig. 13.13):
   When the impulse is given with the operating switch S1, the conveyor belt starts running clockwise. When switch S3 is reached, the heating is switched on. As soon as the limit switch S4 makes contact, the belt stops. The dwell time in the oven is 10 s. After that the part is moved back again until reaching the limit switch S2. The system should be stopped at any time by the stop switch. In this case, the heating is to be switched off immediately and the part is to be moved back to the starting position.

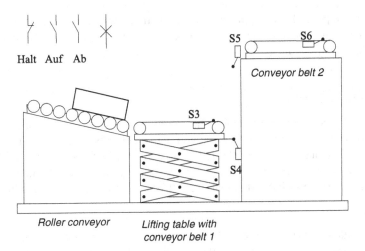

Fig. 13.14 Technology diagram for Exercise 13.31 (EX02) and for Exercise 13.32

Exercise 13.31 (EX02)

Pallet lifting table 1 (Fig. 13.14).
  The lifting table is to be used to transfer pallets to a higher loading platform.

1. When the pushbutton S1 is actuated, the conveyor belt 1 of the lift table is switched on. The pallet rolls over the inclined roller conveyor onto the running conveyor belt 1. By actuating the limit switch S3, the conveyor belt 1 is switched off and the motor for the upward movement of the lifting table is switched on. If the table actuates the limit switch S5, the motor is switched off and both conveyor belts 1 and 2 must run until the limit switch S6 is actuated by the pallet. Now the lift table moves downwards again until S4 signals that the lower limit position has been reached.
2. It must be possible to switch off the system at any time using the pushbutton switch S0. After actuation of S1 the started sequence of the plant shall be continued.
3. As long as the system does not assume the initial position (lifting table down, conveyor belt switched off), the H1 signal lamp should light up.

Exercise 13.32 (EX03)

Pallet lift table 2 (Fig. 13.14).
  In addition to the previous task 13.31, empty pallets are now to be transported back. If an empty pallet is on the conveyor belt 2 (limit switch S6 actuated) and the pushbutton switch S2 is pressed, the upward movement of the lift table is switched on. In the upper end position (S5), both conveyor belts are switched on in counterclockwise rotation. If the limit switch S3 is actuated by the pallet and then released again, the conveyor belts are switched off and the table moves downwards to the lower position.

It must be possible to stop the system at any time by pressing pushbutton S0. After subsequent actuation of the pushbutton S1 or S2, the pallet transport is to be continued in the desired direction.

Exercise 13.33 (EX04)

A transport section, e.g. a conveyor belt, is to be monitored to ensure that no more than 12 and no fewer than 8 parts are present within the section between photoelectric sensor LS1 and photoelectric sensor LS2. If 12 parts are within this range, the stopper shall stop the parts supply. If there are less than 8 parts, a warning signal (lamp or horn) shall be given. After the switch-off signal, the belt should run empty for some time before it stops.

Notice:

Each inflowing part passing the light barrier LS1 shall increase the counter value by 1, and each outflowing part passing the light barrier LS2 shall decrease 1.

Solve this task with PLC! To do this, create the assignment lists, the function block diagram, the instruction list and test your program.

Exercise 13.34 (EX05)

In a floppy disk factory, 10 floppy disks are to be packed in each box. Design a control system! Create the allocation list, the logic diagram, program and test your solution.

Exercise 13.35 (EX06)

In a retail store, a control light in the office should light up when one or more customers are in the store. Design a solution with PLC.

Exercise 13.36 (EX07)

Program a generator that outputs a pulse train at output A0.1 that gives a total duration $T_1 + T_2 = 5s$ and a duty cycle $k = 3/5$. What is the frequency of the generator?

Exercise 13.37 (EX08)

A single-track railway line is divided into four track sections. In order to prevent the simultaneous use of track sections by several trains, a signal is set up at the beginning of each section. In an interlocking, an alarm device shall be triggered when signal 1 and 2 indicate free running or 2 and 3 indicate free running or 3 and 4 indicate free running.

Determine the complete function table for this control.

Fig. 13.15 For Exercise 13.42

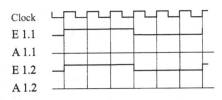

Exercise 13.38 (EX09)

An airlock has three doors. Two immediately following doors must always be closed. The individual doors are opened via door openers. Limit switches report whether the doors are closed. Set up the function table and implement the control with PLC.

Exercise 13.39 (EX10a)

In a large detached house with decentralised hot water supply, five instantaneous water heaters are installed. Due to the high connected load of the instantaneous water heaters, only a maximum of two may be in operation at the same time. The operating status of the instantaneous water heaters is indicated by load shedding relays.

Design an interlock circuit 2 out of 5 by setting up an assignment table and a function table. Then determine the disjunctive normal form from the function table. Implement the circuit with PLC!

Exercise 13.40 (EX10b)

As Exercise 13.39, but four heaters have a connected load of 1 kW each, and the fifth has a load of 2 kW. Implement a circuit which gives an alarm when the power of 2 kW is exceeded!

Exercise 13.41 (EX11)

The note-taking light in a demonstration room may only be lit when the main light is switched off and the on switch is operated.

Exercise 13.42 (EX12)

1. Implement a blinker on output A 1.1, which is stopped with a 0 signal on input E 1.1. (Fig. 13.15).
2. Realize a blinker on output A 1.2, which is stopped with a 1 signal on input E 1.2.

Exercise 13.43 (EX13)

1. Write a PLC program that queries the keys at %ix0.7, %ix0.6 and %ix0.0. The key at %ix0.0 fills the boiler as long as it is pressed, the key at %ix0.6 empties the boiler. The key at %ix0.7 is used for heating.

2. With the key at %ix0.7, the heating is now to be switched on permanently by briefly pressing the key, but only if the boiler is not empty! The key at %ix0.6 switches the heating off.

3. Extension of the task: after reaching $T = 50°$ C the heating should switch off until the boiler has cooled down to 40° C. Now the heating must switch on again and so on until the process is completely ended by means of the key at %ix0.6.

Exercise 13.44 (EX14)

1. Write a PLC program that queries the button at %ix0.7. As long as the button is pressed, the control lamp at %qx0.0 should light up, the heating in the boiler process should heat up and the boiler should be filled via the filling valve.

2. Now each keystroke is to be counted. The number is to be output on the byte %qb3.

3. A second counter is now to indicate how often the temperature of has been reached 40° C or exceeded. This number is to be output on a byte at %qb2.

4. When the temperature has been reached three times, open the outlet valve accumulating.

Exercise 13.45 (EX15)

1. Write a PLC program that queries the button at %ix0.7. As long as the button is pressed, the control lamp at %qx0.0 should light up and the boiler should be filled via the filling valve.

2. Now the heating process is to be started at the push of a button. As soon as are reached 50° C, the heating stops.

3. A counter is now to indicate how often the temperature of has been reached 50° C. This number is to be output on a byte at %qb3.

4. When the temperature has been reached three times 50° C, open the outlet valve accumulating as soon as the temperature has cooled down.

Exercise 13.46 (EX16)

In a production line Fig. 13.16, the manufactured parts must be distributed to three different workstations for further processing.

A light barrier is installed in the conveyor belt, which gives a 0 signal when a part passes through. This signal causes the magnetic actuator to receive an impulse lasting half a second, which causes the switch to move one step further, thus clearing the way to the corresponding station. After the third station, the first station is to be started again. This is achieved by a three-second impulse to the solenoid actuator.

Fig. 13.16 For Exercise 13.46

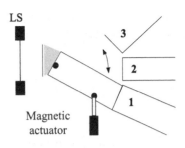

---

1. A measuring vessel is to be filled. A (short) press on the start button starts the filling process. The filling valve should remain open for 3 s. The limit value transmitter is not yet to be taken into account.
2. A (short) press on the stop button also closes the valve before the time from task 1 has elapsed. Likewise, reaching LIS2 closes the filling valve; i.e. after pressing the start button, the filling valve should now be closed after 3 s or by pressing the stop button or by reaching LIS2.
3. As soon as the upper limit value LIS2 is reached, the drain valve should open automatically until LIS1 signals drain. Then the drain valve closes again. During this time, filling should also not be possible.
4. After LIS1 has signalled draining, the valve should not be closed immediately, but only after 3 s, so that the boiler can drain completely.
5. The number of fillings should be displayed with LEDs dual.

Sorting plant 1 (Fig. 13.17).

Wooden beams are transported over a roller system and are to be counted. The beams are of different lengths. Therefore, short and long beams are to be recorded separately. An encoder0 emits a pulse when a beam passes through. If the beam is short, only encoder1 also emits a '1' signal; if the beam is long, both encoder1 and encoder2 emit a '1' signal.

1. Write a PLC program that displays the number of bars in a single-digit 7-segment display. Simulate the encoders with the standard IO process. Initially, the display does not need to go beyond '9', so only a maximum of nine are initially recorded.
2. Now a distinction is to be made between the short and long bars. With two separate counters the respective numbers are to be represented at two displays. (Again only up to max. 9 pieces)

Fig. 13.17 For Exercise 13.48 and for Exercise 13.49

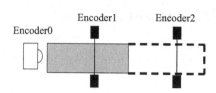

3. If no other encoder or only encoder 2 gives a signal when encoder 0 is pulsed, some kind of fault has occurred. This fault should be signalled by a lamp until an acknowledgement button has been pressed. During the fault indication no counting pulses may reach the counters.
4. In a HEX display, the numbers from 10_{dec} are displayed as 'A', 'B' .... Create a decimal display with two counter components and two 7-segment displays, which displays two digits from '00' ... '09', '10', ... '99'.

Exercise 13.49 (EX19)

Sorting plant 2 (Fig. 13.17).

Wooden beams are transported over a roller system and are to be counted. The beams are of different lengths. Therefore, short and long beams are to be recorded separately. An encoder0 emits a pulse when a beam passes through. If the beam is short, only encoder1 also emits a '1' signal; if the beam is long, both encoder1 and encoder2 emit a '1' signal.

1. Write a PLC program that displays the total number of bars (without distinguishing between long and short) in a single-digit 7-segment display. Simulate the encoders with the standard IO process. The display does not need to be decimal; in this case you can simply use the HEX-Output display for display.
2. Now a distinction is to be made between the short and long bars. With two separate counters, the respective numbers are to be displayed on two displays.
3. If no further encoder or only encoder 2 gives a signal when encoder 0 is pulsed, some kind of fault has occurred. This fault should be signalled by a lamp until an acknowledgement button has been pressed.
4. During the fault indication and for 3 s after acknowledgement, no counting pulses may reach the counters.

Exercise 13.50 (EX20)

In a traffic light control system, the traffic light for the side street should not turn "green" until four cars are waiting. Even if there is less traffic, no car should wait longer than 10 s. The secondary traffic light should show green for 10 s and then switch back to red.

Example: Control of a Mountain Railway

14

Abstract

In this chapter we have worked on the example 4.2.5 (page 116) and 4.4.5 (page 163) from the book by John and Tiegelkamp (Book: John, Karl-Heinz, Tiegelkamp, Michael.: SPS-Programmierung mit IEC 61131-3 – Konzepte und Programmiersprachen, Anforderungen an Programmiersysteme, Entscheidungshilfen, Springer, Heidelberg (2009)) for the programming of a control of a mountain railway – cable car for "PLC-lite" as an exercise and repetition task and commented on it in detail. Beyond the given example, the functionality was extended, e.g. by a demand stop request. With "PLC-lite" this exercise can be carried out on the PC. During the simulation with the program "PLC-lite" the gondola moves in the landscape. By means of control panels the movement of the gondola can be controlled on the screen. In this way, you can safely test all control situations. If a wrong run occurs, the error is clearly displayed.

14.1 Control of a Cable Car

Figure 14.1[1] shows the topography image of the Ifinger cable car ("Bergbahn" in German) "Merano 2000". Located near Merano, Italy, this 3800-m-long cable car can transport over 800 people per hour from the lower station ("Talstation" in German) to the top station ("Bergstation" in German) 1240 m higher at the foot of the Ifinger. This cable car has two gondolas, with only one of them stopping on demand at the middle station ("Mittelstation" in German) near Gsteier. Considering only this one gondola, this corresponds exactly to the example from the book [1], which performs the following functionality:

---

[1] Photo with permission of Geo Marketing GmbH/srl, 39100 Bolzano, Italy.

© The Author(s), under exclusive license to Springer-Verlag GmbH, DE, part of
Springer Nature 2022
H.-J. Adam, M. Adam, *PLC Programming in Instruction list according to IEC
61131-3*, https://doi.org/10.1007/978-3-662-65254-1_14

Fig. 14.1 Ifinger mountain railway

Task

The control system must take the following requirements into account:

- The sensors S1, S2, S3 report TRUE (1) when the gondola is in one of the stations.
  In the counter StationStopp the total sum of all station entries is to be registered.
- The motor for nacelle movement knows the input variables:
  Direction: Forward (TRUE)/Backward (FALSE) (buffered declared)
  StartStop: Start (TRUE), Stop (FALSE)

- There is a DoorOpen switch in the gondola. Switch to 1 means: "Open door", FALSE means: "Close door".
- The motor for door control has two actuators ActOpenDoor and ActCloseDoor (active: on edge from FALSE to TRUE), which cause it to open and close the door respectively.
- A Startbutton sets the gondolas in operation;
  The end of operation can be initiated with the Stop button.
- A warning signal is to be activated in the time between switching off the cable car and the restart.

The PLC project consists of a main program and a function block. In the main program, mainly the input variables (sensors, switches) are queried and the output variables (actuators, displays) are controlled. The functionality is mainly implemented in the function block.

In Exercise 14.1 you can experiment with a cable car control. At first, do not pay attention to the program codes, but "play"! We will explain the code in detail in a moment.

Exercise 14.1 (CC01a)

Start with PLC-lite the project CC01a.plp and the visualization of the process "Mountain railway" (Bergbahn). You can additionally display the process "Standard-I/O 16bit", so that you can better observe the signals of the process.

Start the program with "Run".

After pressing the green start button on the *"main panel"* at the top left (which is located in the cable car control room), the cabin door closes after a few seconds and the gondola starts moving. It stops at the mid station, continues after a short stop, reverses direction after stopping at the top station and returns to the valley.

In the simulation with PLC-lite, the small *"cabin panel"* with the green door button and the red stop button travels with the cabin. For space reasons it is not "mounted" inside but outside next to the cabin. In addition, there is a "copy" of this panel in the upper left corner, because otherwise it is difficult to click the buttons with the mouse while driving. With the button "Door open" you can open the closing door once more, to allow e.g. late arriving passengers to enter. The red stop button on the cabin panel is still without function, it will be put into operation in Exercise 14.7.

With the red stop button (end of operation, power off) on the main panel you give the signal to the gondola to continue driving and to stop at the <u>next stop at the lower station</u> until you press the start button again. Coloured rings around the buttons indicate the operating states Ready, Run, End of operation requested and End of stop reached (Power off).

At the middle station at Gsteier there is another small *"station panel"*, with a display of the direction of travel and two buttons for a stop request up or down. These are not yet used in the first example, but will be programmed in Exercise 14.10.

In addition, there is a *"revision panel"* in the upper left corner with four buttons: fault acknowledgement, open door (directly, without PLC program), drive up or down (also directly). Attention: You can use these buttons to open the door during the run (fault simulation) and run the gondola against the ends, beyond the normal end points.

We now look at the program code of the example in detail.

14.2   Cable Car: Main Programme

First the input variables for the input buttons and sensors, the output variables for the displays of some process states and for the actuators and other variables are declared. The displays for direction (Richtung), moving (Fahrt), station stops (Start/Stop), door open and door close (Tür öffnen, Tür zu), wait (warten) and the displays for the operating states ready, travel mode, end of operation requested and final stop reached (PowerOff) are set independently by the simulation process and do not have to be programmed separately.

```
PROGRAM CC01a
(* Processes:            *)
(*  Bergbahn (cable car) *)
(* example from "SPS-Programmierung mit IEC 61131-3" *)
(* (c) 2008 John/Tiegelkamp (Springer-Verlag)        *)


VAR
Start AT              %I0.0: BOOL (* 1 = Start *)
End AT                %I0.1: BOOL (* 1 = Stop in lower station *)
ButtonOpenDoor AT     %I0.3: BOOL (* 1 = open *)
SensorDoorClosed AT   %I0.4: BOOL (* 1 = closed *)
LowerStation AT       %I1.0: BOOL (* 1 = in station *)
MiddleStation AT      %I1.1: BOOL
TopStation AT         %I1.2: BOOL
Action AT             %Q0.0: BOOL (* 1 = Plant on *)
ActOpenDoor AT        %Q0.1: BOOL (* 1 = Open door *)
ActCloseDoor AT       %Q0.2: BOOL (* 1 = close *)
PowerOff AT           %Q0.3: BOOL (* 1 = stop, cabin in lower station *)
StartStop AT          %Q0.4: BOOL (* 1 = drive *)
MoveDown AT           %Q0.5: BOOL (* 1 = downwards *)
DisplayEnd AT         %Q0.6: BOOL
```

```
Warning AT              %Q0.7: BOOL
StationStopp AT         %QB3:  INT
DisplayWait AT          %Q1.7: BOOL
cycle1:                        BOOL
CabinControl:                  CCControl
END_VAR
```

```
(* set initial values *)
LDN    cycle1
S      PowerOff
S      DisplayEnd
S      cycle1
```

Evaluation of the start or stop button and the PowerOff signal:

```
(* Load physical I/O values (sensor information) *)
```

```
LD     Start
ST     CabinControl.CablewayOn
LD     PowerOff
ANDN   Start
JMPC   NoOperation
LD     End
ST     CabinControl.CablewayOff
```

Determine station stays and transfer them to the "CabinControl" control function block
BBControl:

```
(* stops *)
LD     lowerStation
ST     CabinControl.St1
LD     MiddleStation
ST     CabinControl.St2
LD     TopStation
ST     CabinControl.St3
```

```
Activate:
(* Activate mountain railway control FB *)
LD     ButtonOpenDoor
ST     CabinControl.ButtonOpenDoor
LD     StartStop
```

```
ST    CabinControl.CabinRun
CAL   CabinControl
```

Read values from the control function module and pass them to variables in the main program:

```
(* Store physical I/O values (Motor Information) *)
LD    CabinControl.ActOpenDoor
ST    ActOpenDoor
LD    CabinControl.EndSignal
ST    PowerOff
// because in PLC-lite no type VAR_IN_OUT ->
LD    CabinControl.CabinRunOut
AND   SensorDoorClosed
ST    StartStop
LD    CabinControl.DirectionOut
ST    MoveDown
LD    CabinControl.PowerOffOut
ST    DisplayEnd
// additional control displays:
LD    CabinControl.StationStoppOut
ST    StationStopp
LD    CabinControl.Wait
ST    DisplayWait
LD    CabinControl.Action
ST    Action
JMP   End_POU
```

Gondola is in the lower station and operation is stopped (PowerOff):

```
NoOperation:
LD    FALSE
ST    ActCloseDoor
STN   ActOpenDoor
ST    CabinControl.CablewayOn
ST    StartStop
ST    CabinControl.CabinRun
CAL   CabinControl
// because in PLC-lite no type VAR_IN_OUT ->
LD    CabinControl.Action
ST    Action
```

```
End_POU:
END_PROGRAM
```

14.3   Edge Detection

The system has a sensor for each stop station, which indicates the stay of the gondola in the respective station by a 1. When the car enters the station, the sensor signal goes from 0 to 1. This "rising edge" indicates that the car has just entered the station. In [1], variables of type BOOL with the attribute R_EDGE are therefore used for the sensor signal. This attribute is not implemented in PLC-lite, but a standard function block R_TRIG is. With this, edge detection is also possible.

The function block R_TRIG is the standard specified in [4] and [6][2]: These blocks must operate according to the following rules.

- In the case of a function module R_TRIG, after the transition of the input CLK from 0 to 1, the output Q goes to the Boolean value 1 during the subsequent first execution (CAL ...) of the function module and returns to 0 during the second execution.
  The same applies to F_TRIG: after the transition of input CLK from 1 to 0, output Q remains at the value 1 from the next to the next but one execution of the function block.
- NOTE: If the input CLK of an instance of type R_TRIG is connected to a value 1, its output Q will go to 1 and remain so after its first execution following a cold start. From the next and all subsequent executions, output Q will go to 0 and remain so.
  The same applies to an instance F_TRIG, whose input CLK is unconnected or is connected with the value FALSE.

Rising Edge Detection
In the following function module you can see the IL with which the above rules can be implemented. You do not need to program this! This standard function block is integrated in PLC-lite. This program code only serves as an illustration for you if you want to understand how it works.

```
FUNCTION_BLOCK BOOL_R_TRIG
  VAR_INPUT
    CLOCK: BOOL
  END_VAR
```

---

[2] German: [4] EN 61131-3:2003 (Kapitel 2.5.2.3.2 Flankenerkennung / Tabelle 35 – Standard-Funktionsbausteine Flankenerkennung) or English: [6] E DIN IEC 61131-3:2009-12 – Draft – (6.5.3.5.2 Bistable elements/Table 41 – Standard bistable function blocks).

```
VAR_OUTPUT
   EDGE: BOOL
END_VAR
VAR
   M: BOOL
END_VAR
LD CLOCK
ANDN M
ST EDGE
LD CLOCK
ST M
END_FUNCTION_BLOCK
```

Exercise 14.2 (FLASH141)

In Sect. 10.3 you programmed a running light that required an initial value to be set. Compare the solution from there with the method given here, which uses the function block F_TRIG.

The function block Cycle1: F_TRIG is called exactly once at the beginning of each cycle with CAL Cycle1. Output Cycle1.Q therefore remains at the value 1 from the start of the program for exactly the first cycle and goes constantly to 0 for the rest of the program run. Note that you do not have to assign a value to input Cycle1.CLK.

Create the program for the running light and test it.

```
Program FlashLightFlash141
(*Processes: *)
(* standard I/O byte 1 *)
(* Hex-Output byte 1    *)
var
   Pulse: TON
   Cycle1: F_TRIG
   Bit0 AT %QX1.0: BOOL
   Bit7 AT %QX1.7: BOOL
   Byte1 AT %QB1: USINT
end_var
(* set bit *)
   CAL Cycle1
   LD Cycle1.Q
   S Bit0

   ...
```

14.4   **Cable Car (Bergbahn): Control Function Block**

We can now look at the control function block.

First, the variables are defined:

```
FUNCTION_BLOCK CCCONTROL
VAR_INPUT
  CablewayOn:      BOOL //R_EDGE (* key to start *)
  CablewayOff:      BOOL          (* initiate end *)
  St1,St2,St3:     BOOL //R_EDGE (* sensor station *)
  ButtonOpenDoor:  BOOL          (* 1 = open, 0 = close *)
END_VAR

VAR_INPUT //_IN_OUT
  CabinRun:        BOOL          (* 1 = car moves *)
END_VAR

VAR_OUTPUT
  ActOpenDoor:      BOOL
  ActCloseDoor:     BOOL
  CabinRunOut:      BOOL
  DirectionOut :    BOOL
  PowerOffOut:      BOOL
  Action:           BOOL
  StationStoppOut: INT
  Wait:             BOOL
END_VAR

VAR_OUTPUT // RETAIN
  EndSignal:        BOOL         (* Warning signal PowerOff *)
END_VAR

VAR
  StationStopp:     CTU          (* station entrances *)
  DoorTime:         TON
  Entry:            BOOL
  StartPuls:        R_TRIG
  St1Puls:          R_TRIG
  St2Puls:          R_TRIG
  St3Puls:          R_TRIG
END_VAR

VAR // RETAIN
  Direction:        BOOL
  DirectionFF:      RS
```

```
END_VAR
```

Now the function blocks for edge detection are prepared:

```
LD      CablewayOff
S       PowerOffOut

LD      CablewayOn
ST      StartPuls.CLK
CAL     StartPuls
LD St3
ST St3Puls.CLK
CAL ST3Pulse
LD St2
ST St2Puls.CLK
CAL ST2Pulse
LD St1
ST St1Puls.CLK
CAL ST1Pulse
```

Exercise 14.1 requires a warning signal to be output between switching off the path and restarting it. This end signal remains because of the declaration with the attribute RETAIN (also in case of power failure). Therefore, at program start the buffered value and not automatically the initial value 0 is assigned. This is meaningless here with PLC-lite because the RETAIN attribute is not available. However, we leave the reset in the program as it is carried out in [1].

```
LD      StartPuls.Q     (* the first call? *)
R       EndSignal       (* because of attribute RETAIN *)
S       Action
JMPC    ResetCounter
JMP     GoEntry

ResetCounter:
LD      1
ST      StationStop.R
CAL     StationStop
LD      1
ST      StationStop.CU
LD      0
ST      StationStop.R
CAL     StationStop
LD      StationStop.CV
ST      StationStopOut
JMP     CabinClose
```

After entering a station, several actions are required: stop the engine, open the door and count up the station counter.

```
GoEntry:
LD      St1Puls.Q
OR      St2Puls.Q
OR      St3Puls.Q
ST      Entry
R       CabinRun
S       ActOpenDoor
ST      StationStop.CU
LD      0
ST      StationStop.R
CAL     StationStop
LD      StationStop.CV
ST      StationStopOut
```

Now we check whether the car is in the lower station (St1) or the top station (St3). We do this with XOR: in case of a sensor error it could happen that both sensors indicate 1. We should still catch this error case and handle it accordingly (e.g. initiate an emergency stop). Also for safety reasons, we do not simply switch the direction by the command sequence LD direction STN direction, but use the flip-flop directionFF. To set the new direction we use the static signals of the stations.

```
(* possibly change direction *)
LD    St1             (* static signals!   *)
XOR   St3
JMPCN NoSwitch        (* no 1? -> continue *)
(*  change direction *)
LD    St1             (* static signals!   *)
ST    DirectionFF.R1
LD    St3
ST    DirectionFF.S
CAL   DirectionFF
LD    DirectionFF.Q1
ST    DirectionOut
```

The car door may only be opened when staying in a station. The signal from the door open button is linked accordingly:

```
NoSwitch:
LD    ButtonOpenDoor  (* Query door switch *)
AND(  St1             (* when staying in station *)
OR    St2
```

```
OR      St3
)
S       ActOpenDoor
```

If the end of service has been requested and we have pulled into the lower station, then we are now done …

```
(* End + in station -> POU end *)
LD      PowerOffOut    (* circuit breaker actuated? *)
AND     St1Puls.Q      (* entry in lower station    *)
S       EndSignal
R       Action
JMPC    PoeEnd


CabinClose:
LD      ActOpenDoor
STN     ActCloseDoor   (* never both the same! *)
(* Continue after 4 sec. *)
LDN     ButtonOpenDoor
ANDN    CabinRun
ST      DoorTime.IN
S       Wait           (* control display *)
LD      T#4s
ST      DoorTime.PT
CAL     DoorTime
LD      DoorTime.Q     (* Time expired? *)
AND     Action
S       CabinRun
R       ActOpenDoor
R       Wait
```

End of the function module and return to the calling main program:

```
PoeEnd:
LD      CabinRun
ST      CabinRunOut
END_FUNCTION_BLOCK
```

Exercise 14.3 (CC01FF)

For example, using a function table, justify why the flip-flop switches `directionFF` to reverse direction only in the two end stations.

Specify a circuit that detects the error case: both sensors give signal and turns on a warning light.

Exercise 14.4 (CC01b)

Our display of station entries has the "flaw" that the numbers are displayed in HEX code. To display the numbers in BCD code, we have developed a function in Exercise 11.4 (BCD111.PLP):

```
Function hex_to_bcd: Byte (f_h2bcd.il)
```

Extend the mountain railway Exercise 14.1 by this function to display the station entries in BCD notation!

Exercise 14.5 (CC01c)

The display should now not show the number of station entries, but the respective station number for a station stop: for the lower station a 1, for the middle station the number 2 and for the top station the 3. During the journey the character string "FA" should be displayed.

Solution hint: This is a code converter that must convert the 1-out-of-3 code into a dual code. In Sects. 2.14 and 2.17, you created a circuit equation by evaluating the minterms ("AND-before-OR") of a function table. We can proceed similarly here. We have the three input signals lowerStation, MiddleStation and TopStation, of which either none or only one is "1". As outputs of the function table we provide the two bits for 2^1 and 2^0 of the station number. These are combined to give the station number as a numerical value 01^{BCD}, 10^{BCD} or 11^{BCD} (as decimal numbers: 1, 2 or 3).

Change the mountain railway Exercise 14.1 to show the station number instead of the number of station entries.

Display of the Station Number

The solution of Exercise 14.5 can be realized with the following program part: First we determine the more significant bit 2^1. It must be 1, if the car is in the middle station or in the top station. Here we specify the complete minterms, although this would not be necessary, because the car is always only in one station. But: if a sensor error would occur, this situation can be recognized and appropriate measures can be taken. The minterms result in values of type BOOL, which must be converted into numerical values (type INT) for further processing. In Sect. 8.2 we already dealt with type conversions. The result is temporarily stored in the variable station number:

```
(* Output station number as a number *)
LD MiddleStation (* bit 1 determine *)
ANDN lowerStation
ANDN TopStation
OR( TopStation
ANDN MiddleStation
```

```
ANDN BotomStation
)
BOOL_TO_INT
MUL 2 (* push left *)
ST StationNumber
```

Determination of the least significant bit 2^0, also the complete minterms. This bit must be 1 if the car is in the lower station or the top station:

```
LD lowerStation (* bit 0 determine *)
ANDN MiddleStation
ANDN TopStation
OR( TopStation
ANDN MiddleStation
ANDN lowerStation
)
BOOL_TO_INT
```

Now the variable StationNumber is added to this intermediate result and the sum is stored back as the final result in the variable StationNumber:

```
ADD StationNumber (* combine *)
ST StationNumber
```

StationNumber is 0 if the car is not in a station. At this point one could also evaluate the variable StartStop:

```
EQ 0
JMPC DisplayRide (* not in station *)
LD StationNumber
JMP DisplayStation
DisplayRide:
LD 16#FA
DisplayStation:
ST StationStop (* show *)
```

Exercise 14.6 (CC01d)

Modify the solution from Exercise 14.4 by using the variable StartStop as a condition for the ride display. Discuss the consequences of the two different methods.

14.5   Mountain Railway with Stop Request from the Cabin

The mountain railway is now to stop at the Gsteier middle station only on request. Initially, this should only be possible with the button in the cabin panel in the cabin. This means that only persons travelling with the cabin can initiate the on-demand stop.

Exercise 14.7 (CC02a)

Extend the program from Exercise 14.1 by the request of the stop button on the car panel and the car stop at the middle station, if the stop button was pressed before.

Solution Tip

A variable is set to store the keystroke on the hold button. Only when this variable is set is the sensor signal passed on to the control function block. This stop request variable is reset when the exit is made. To register the exit, you can use a variable of type F_TRIG.

The button and the display are connected to the following inputs and outputs respectively:

```
StopSt2CabButton AT %I1.3: BOOL
StopSt2Cabin AT %Q1.3: BOOL
```

You do not need to make any changes in the control function module, only in the main program.

Stop at the Middle Station at Gsteier Only on Request

The control for Exercise 14.7 can be realized with the following program part:

```
LD StopSt2CabButton
S StopSt2Cabin (* Register Halt *)
LD MiddleStation
AND StopSt2Cabin (* Stop on request *)
ST CableControl.St2
ST MiddleStationExit.CLK
CAL MiddleStationExit
LD MiddleStationExit.Q
R StopSt2Cabin (* Delete request *)
```

Exercise 14.8 (CC02b)

Now the display of the station stops should be shown as BCD numbers and not as HEX numbers as in Exercise 14.4.

Exercise 14.9 (CC02c)

As in Exercise 14.5, the station numbers should now be displayed during the stays.

14.6   Cable Car with Stop Request from the Middle Station

Now, at the Gsteier mid station, the up/down buttons on the station panel at the station are also to be put into operation so that a hiker can stop the cabin to board. In order to avoid unnecessary stops, there are different buttons for the stop request for uphill or downhill travel. The stop request should therefore be executed depending on the direction.

Exercise 14.10 (CC03a)

Extend the program from Exercise 14.9 by the request of the stop button on the station panel and the car stop at the mid station, if the stop button "up" or "down" was pressed before and the car moves in the corresponding direction. If the car travels through the mid station in the direction not requested, the request remains set until the car travels in the requested direction on its way back.

The buttons and the displays are connected to the following inputs and outputs:

```
RequestUpButton AT %I1.4: BOOL
RequestUp AT %Q1.4: BOOL
RequestDownButton AT %I1.5: BOOL
RequestDown AT %Q1.5: BOOL
```

Solution hint: The condition for the stop request is now:

$$MiddleStation \;\wedge$$
$$(StopSt2Cabin \;\vee$$
$$\left(RequestOn \wedge \overline{TravelDirection}\right) \;\vee$$
$$\left(RequestFrom \wedge TravelDirection\right)$$
$$)$$

The following conditions apply to resetting the stop request:
Reset `StopSt2Cabin` as in Exercise 14.7:

$$MiddleStationExit.Q$$

Reset `RequestUp`:

$$MiddleStationExit.Q \wedge RequestOn \wedge \overline{TravelDirection})$$

Reset `RequestSt2Down`:
$$MiddleStationExit.\,Q \wedge RequestFrom \wedge TravelDirection)$$

Demand Stop at Gsteier Depending on Direction

The control for Exercise 14.10 can be realized with the following program part:

First, the corresponding keystrokes must be queried and stored.

```
LD StopSt2CabButton
S StopSt2Cabin (* Register stop *)
LD RequestUpButton
S RequestUp (* Register stop *)
LD RequestDownButton
S RequestDown (* Register stop *)
```

We now check whether there is a condition for a stop. When entering the middle station we check if the stop button from the cabin is pressed or if the stop button for uphill was pressed at the station panel and the car is on uphill or if the stop button for downhill was pressed and the car is on downhill.

```
LD MiddleStation
AND( StopSt2Cabin (* Stop request *)
  OR( RequestUp (* only for uphill *)
    ANDN Direction
  )
  OR( RequestDown (* only for downhill *)
    AND Direction
```

If the car has entered the middle station and a stop condition exists, the previous expression results in 1, St2 is activated in the controller and the car stops. In addition, the variable MiddleStationExit is triggered. It is of type F_TRIG, the output Q remains 0 until the car has left the middle station. Then the middle station becomes 0 again, and the 1 to 0 transition (negative edge!) at the clock input CLK causes output Q to become 1 for one program cycle. Exactly in this cycle the requests to be cleared are reset.

```
ST    CableControl.St2
ST    MiddleStationExit.CLK
CAL   MiddleStationExit
LD    MiddleStationExit.Q (* Exit? *)
AND   StopSt2Cabin
R     StopSt2Cabin (* Delete request *)
LD    MiddleStationExit.Q
AND   RequestUp
ANDN  Direction
R     RequestUp (* Delete request *)
LD    MiddleStationExit.Q
AND   RequestDown
AND   Direction
R     RequestDown (* Delete request *)
```

14.7   Treatment of Operational Faults

Below the station names you will see fields "Sensor ok". If you click on it, you simulate the malfunctions "Cable break" (i.e. the sensor permanently gives 0-signal) and "Blocking" (i.e. the sensor permanently gives 1-signal).

Exercise 14.11 (CC04a)

If you run the program from Exercise 14.1, you can observe, among other things, how each "wrong click" on a station sensor increases the number of entries.

You can also see what happens when the gondola moves beyond the end stops! Also use the buttons for the up and down movement of the gondola on the revision panel.

Test the programs extensively to detect as many error situations as possible!

Error Handling Routines

You can now include your own error handling. If, for example, more than one station sensor gives a signal at the same time, a jump is made to the "ErrorAction" marker in the following code fragment. There you can program a reaction to this error cause.

See also your solution to Exercise 14.3.

```
LDN     lowerStation
ANDN    MiddleStation
ANDN    TopStation
OR      ( lowerStation
ANDN    MiddleStation
ANDN    TopStation
)
OR      ( MiddleStation
ANDN    lowerStation
ANDN    TopStation
)
OR      ( TopStation
ANDN lowerStation
ANDN MiddleStation
)
JMPCN   Erroraction
JMP     NoFault
ErrorAction:
LD      TRUE
S       Warning
jmp     NoOperation
NoFault:
LD      TRUE
R       Warning
```

Emergency Operation

Furthermore, there are additional sensors in the system as limit switches, which emit a 1-signal when not actuated. These signals come a little later than the normal stop sensors. The status of these sensors is displayed on the revision panel to the right of the buttons for upward or downward travel. The signals are connected to the inputs %I2.0, %I2.1 and %I2.2.

```
lowerStation2 AT %I2.0: BOOL (* 0=end stop *)
MiddleStation2 AT %I2.1: BOOL (* 0=hold point *)
TopStation2 AT %I2.2: BOOL (* 0=end stop *)
```

You can use these sensors for emergency operation.

Exercise 14.12 (CC04b)

Extend the program from Exercise 14.10 for emergency operation. The error handling should also evaluate the limit switches. Control the warning in emergency operation (%Q0.7)! Test the program extensively to detect all error situations if possible.

Structure and Programming of a PLC

15

Abstract

In the case of a *hardwired* control system, the program is determined by the fixed wiring of the active switching elements (gates) and can only be changed by changing the wire connections, which is usually very time-consuming.

In a *stored-program* control system, the signal processing is performed by a computer. The instructions to the computer necessary for the program realization of the control are stored in a memory. The totality of these instructions is referred to as a *"program"*.

If the program is stored in a read-only memory (ROM), the control program can only be changed by replacing this memory (chip); this is then a replaceable programmable controller. A freely programmable controller uses a random access memory (RAM) as program memory. Here the program of the control can be easily changed by reprogramming the memory.

In principle, all programmable logic controllers have the same structure as any computer: central processing unit, memory and input and output units. The central processing unit and memory usually form a single unit and are referred to as the automation device.

Figure 15.1 shows an overview of the control types. In the case of a *hardwired-programmed* control, the program is defined by the fixed wiring of the active switching elements (gates) and can only be changed by changing the line connections, which is usually very time-consuming.

H.-J. Adam, M. Adam, *PLC Programming in Instruction list according to IEC
61131-3*, https://doi.org/10.1007/978-3-662-65254-1_15

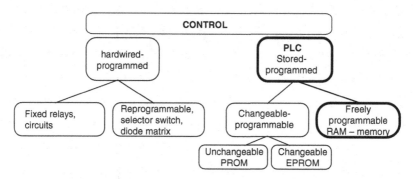

Fig. 15.1 Programmable controllers

In a *stored-program* control system, the signal processing is performed by a computer. The instructions to the computer necessary for the program realization of the control are stored in a memory. The totality of these instructions is referred to as a *"program"*.

If the program is stored in a read-only memory (ROM), the control program can only be changed by replacing this memory (chip); this is then a replaceable programmable controller. A freely programmable controller uses a random access memory (RAM) as program memory. Here the program of the control can be easily changed by reprogramming the memory.

In principle, all programmable logic controllers have the same structure as any computer: central processing unit, memory and input and output units. The central processing unit and memory usually form a single unit and are referred to as the automation device.

Binary and digital signals can be input or output via the input and output units. Analog signals must be converted into digital signals. The corresponding signal converters are called analog-to-digital converters (A/D converters) or digital-to-analog converters (D/A converters).

The creation of a *control program is* done in the following four steps:

1. **Functional program description**
   The control task is described verbally, input and output data are specified. In controllers, input and output data are predominantly binary, digital and analog signals that come from the process (process signals from sensors) or go to the process (commands to actuators).
2. **Create program flow chart (structure chart)**
   The program flow chart and the structure chart are aids that are used to "construct" programs. With the aid of graphic symbols, the sequence of the program is clearly represented. From a rough program overview, the program structure is obtained by refining it step by step: the control task is thus divided into individual steps that can be implemented by the computer. The structure diagram and program flow chart are independent of the programming language used.

3. **Program creation**

The individual steps that can be seen from the structure diagram or program flow chart are transferred to the programming language. The result is, for example, the instruction list, which can be translated into machine code by the compiler or processed directly by the interpreter. Interpreters are often used in the PLC, since program errors can then be found more easily in step mode.

4. **Test and documentation**

Program testing and documentation must be carried out very carefully. All functions of the control must be tested in interaction with the process (on-line), or if this is not possible, by simulation (off-line) under all conceivable operating conditions. Good documentation is required in order to be able to make subsequent changes to the program with reasonable effort.

The process signals are present at the inputs. The program is stored in the user memory and is selected step by step by the control unit. The selected control instruction is transferred to the control unit and processed. In the process, for example, inputs, outputs, flags, times are interrogated for their signal state and linked in order to control outputs, flags, etc. The command is executed in the user memory. After execution of an instruction, the instruction counter is incremented so that the next memory cell is controlled.

After execution of the last instruction, the program is started again at the first address: the program is repeated cyclically.

Standard Compliance of PLC-Lite

16

Abstract

In this chapter you will find references to the standard IEC 61131-3, as far as they are realized in PLC-lite. The language and structure elements of the IEC 61131-3 standard described here document the properties of the "PLC-lite" program.

The following Tables 16.1, 16.2, 16.3, 16.4, 16.5, 16.6, 16.7 and 16.8 describe the properties of PLC-lite with regard to the IEC 61131 standard. This system meets the requirements of IEC 61131-3 in the following properties of the language: IL.

Table 16.1 Text elements

Term	Description
Character set	Only characters (letters, digits and special characters) from the "basic code table" of ISO/IEC-646 may be used. The upper/lower case is not significant for the language elements
Identifier	Identifiers, e.g. variable names, must begin with a letter or an underscore. They must not contain umlauts (äöü) or ß!
Keywords	The keywords are used by the programming system as syntactical elements and can therefore not be used for user-defined identifiers
Space	Spaces are not allowed within identifiers, keywords or literals. However, any number of spaces are allowed in all other places
Comments	Comments are enclosed by the character combinations (* and *). They may be used wherever spaces are allowed. It is not permitted to "nest" comments; e.g. (* comment (* nested *) *)
Literal	Literals are used to represent data values. They are displayed according to the following tables

© The Author(s), under exclusive license to Springer-Verlag GmbH, DE, part of Springer Nature 2022
H.-J. Adam, M. Adam, *PLC Programming in Instruction list according to IEC 61131-3*, https://doi.org/10.1007/978-3-662-65254-1_16

Table 16.2 Numerical literals

Name	Examples	Note
Integer	−12	
	0	
	23	
	+56	
Real numbers	−12.0	n.s.[a]
	0.0	
	3.14159	
Real numbers with exponent	−1.34E-12	n.s.
	1.0e+6	
Binary numbers	2#1111_1111	
	2#11100000	
Octal numbers	8#377	n.s.
	8#340	
Hexadecimal numbers	16#ff	
	16#FF	
	16#E9	
Boolean values	0 or FALSE	
	1 or TRUE	
Literals with type specification	INT#-12	
	BYTE#2#0110_1100	
	UINT#16#EFFE	

[a]n.s. means: in PLC-lite this property is not supported

Table 16.3 String literals

Description
No character strings (texts, strings) can be edited with PLC-lite

Table 16.4 Time literals

Name	Examples	Note
Duration	T#14 ms	
	t#14h12m18s3.5 ms	
	t#14h_12m_18s_3.5 ms	
	Time#14.7 s	
Date	Date#1984-06-25	n.s.[a]
	D#1984-06-25	
Time of day	Time_of_Day#15:36:55.36	n.s.
	TOD#15:36:55.36	
Date and time	Date_and_Time#1984-06-25-15:36:55.36	n.s.
	DT#1984-06-25-15:36:55.36	

[a]n.s. means: in PLC-lite this property is not supported

Table 16.5 Data types

Keyword	Data type	Bits	Area	Initialization value	
BOOL	Logical Boolean	1	0, 1 (FALSE, TRUE)	0	
SINT	Short integer	8	$-128 \dots +127$	0	
INT	Integer	16	$-32768 \dots 32767$	0	
DINT	Double integer	32	$(-2^{31}) \dots (+2^{31}-1)$	0	
LINT	Long integer	64	$(-2^{63}) \dots (+2^{63}-1)$	0	n.s.[a]
USINT	Unsigned short integer	8	$0 \dots 255$	0	
UINT	Unsigned integer	16	$0 \dots 65535$	0	
UDINT	Unsigned double integer	32	$0 \dots (2^{32}-1)$	0	
ULINT	Unsigned long integer	64	$0 \dots (2^{64}-1)$	0	n.s.
REAL	Real number	32	According to IEC 559 for floating point format	0	n.s.
LREAL	Long real number	64	According to IEC 559 for floating point format	0	n.s.
TIME	Duration	32	$0 \dots \pm$ approx. 24 days Smallest unit: 1 ms	T#0 s	
DATE	Date			D#0001-01-01	n.s.
TOD	Time			TOD#00:00:00	n.s.
DT	Date and time			DT#0001-01-01-00:00:00	n.s.
STRING	Variable length string		A numeric value range is not applicable for these data types	"(empty string)"	n.s.
BYTE	Bit sequence	8		0	
WORD	Bit sequence	16		0	
DWORD	Bit sequence	32		0	
LWORD	Bit sequence	64		0	

[a]n.s. means: in PLC-lite this property is not supported

Table 16.6 General data types

The general data types are identified by the prefix 'ANY'				
ANY_ELEMENTARY				
ANY_MAGNITUDE		ANY_BIT	ANY_STRING[a]	ANY_DATE[a]
ANY_NUM				TIME
ANY_INT	ANY_REAL[a]			
LINT[a], DINT, INT, SINT, ULINT[a], UDINT, UINT, USINT	LREAL[a], REAL[a]	LWORD[a], DWORD, WORD, BYTE, BOOL	STRING[a], WSTRING[a]	DATE_AND_TIME[a], DATE, TIME_OF_DAY[a]

[a]In PLC-lite this property is not supported

Table 16.7 Derived data types[a]

Declarations (agreements)

The user can use the textual construction `TYPE ... END_TYPE` to declare data types by derivation from others and thus use further types in addition to the elementary data types

Initialization

The default value is the value of the first identifier of the enumeration list or the lowest value of the range, or it is specified by the assignment operator :=. For strings, the maximum length can be specified in parentheses

[a]The properties "derived data types" are not included in PLC-lite

Table 16.8 Variable

Single-element variable They contain only a single data element. This can be one of the elementary types or a type derived from an elementary type. Single-element variables can be represented directly: The identifier begins with the % character. Subsequent characters specify the location, size, and hierarchical physical or logical address. The inputs of the PLC are memory location I, outputs are memory location Q and flags are memory location M. For the addresses, note that counting starts at 0

Direct variable (examples)	Size, type	Meaning
`%I0.1`	1 bit, Boolean	Input, byte0, bit1
`%I1`	1 bit	Input, byte0, bit1
`%IB1`	1 byte = 8 bit	Input, byte 1
`%IW0`	2 byte = 16 bit, WORD	Input, word 0, consisting of byte 1 and byte 0
`%ID0`	4 bytes = 32 bits, DWORD	Input, double word 0, consisting of bytes 3–0
`%IL0`	8 byte = 64 bit, LWORD[a]	Input, long word 0, consisting of bytes 7–0
`%Q...`		Output
`%M...`		Flag (memory register/bit)

Multi-element variable[b]

Multi-element variables are the fields (ARRAYs) and the structures (STRUCTUREs)

Declaration of variables

The declaration of variables is terminated for each variable type separately by the keyword 'END_VAR'. Examples for the application can be found in the working part of this book. In PLC-lite, only the properties addressed in the exercises and tasks can be used

(continued)

Table 16.8 (continued)

Keyword/designator	Use, scope	Note
VAR	Within the organizational unit	
VAR_INPUT	Values are supplied from outside the POU, cannot be changed within the POU	
VAR_OUTPUT	Values are delivered from POU to the outside	
VAR_IN_OUT	Values are supplied externally, can be changed within the POU, and are returned externally	n.s.[c]
VAR_EXTERNAL	Values are supplied by configuration	n.s.
VAR_GLOBAL	Global variable valid everywhere	n.s.
VAR_ACCESS	Access paths for communication services according to IEC 1131-5	n.s.
END_VAR	Closes variable declarations	
RETAIN	Buffered variable, last value is restored after warm start	n.s.
CONSTANT	Not changeable (=constant)	n.s.
AT	Assigning a specific storage location	

[a]In PLC-lite this property is not supported
[b]Multielement variables are not included in PLC-lite
[c]n.s. means: in PLC-lite this property is not supported

16.1   Program Organisation Units (POU)

The program organisation units (e.g. "main program", function, function module) are described in Tables 16.9 and 16.10.

Functions
A function returns exactly one data element after the call. A field or a structure is also a data element, even though it contains several values. In the statement list language, the return value is in the Current Result (CR).

The Tables 16.9 and 16.10 list the functions and function blocks implemented in PLC-lite.

Standard Function Blocks
Only the standard function blocks implemented in PLC-lite are described in Table 16.10.

Table 16.9 Standard functions

Type conversion functions

Function	Return value, example	Note
`* _TO_ **`	`*`: Input data type, e.g. `INT` `**`: Output data type, e.g. `USINT`	
`BCD_ TO_ **`	Input data type: `ANY_BIT`, BCD format `**`: Output data type, e.g. `INT`	n.s.[a]
`* _TO_BCD`	`*`: Input data type, e.g. `INT` Output data type: `ANY_BIT`, BCD format	n.s.
`TRUNC`	Truncation in direction zero Input type: `ANY_REAL`, Output type: `ANY_INT`	n.s.

Standard arithmetic functions

Data types: `ANY_NUM`, `TIME`

Function	Return value, example	Note
`ADD`	Addition,	Number of inputs
`MUL`	Multiplication	expandable[b]
`SUB`	Subtraction	Two inputs
`DIV`	Division	
`MOD`	Division remainder	

Standard functions for bit sequences

Data types: `ANY_BIT`

Function	Return value, example	Note
`SHL`	Shift left by N bit, right fill with zero	
`SHR`	Shift right by N bit, fill left with zero	
`ROL`	Rotate left by N bit, "in a circle"	
`ROR`	Rotate right by N bit, "in a circle"	
`AND`	Bitwise AND operation	Number of inputs
`OR`	Bitwise OR operation	expandable[b]
`XOR`	Bitwise exclusive OR operation	
`NOT`	Negation	

Standard functions for comparison

Input: all data types (`ANY_BIT` are interpreted as unsigned integers)

Output: `BOOL`

Function	Return value, example	Note
`GT, >`	"Greater than": returns `TRUE`, if operand > `CR`	
`GE, >=`	"Greater or equal": `TRUE` if greater than or equal to `CR`	
`EQ, =`	"Equal": `TRUE` if equal to `CR`	
`NE, <>`	"Not equal": `TRUE` if unequal to `CR`	
`LE, <=`	"Less or equal": `TRUE` if less than or equal to `CR`	
`LT, <`	"Less than": `TRUE` if smaller than to `CR`	
`MAX`		n.s.[a]
`MIN`		n.s.

[a]n.s. means: in PLC-lite this property is not supported

[b]Expandability of the inputs is not supported in PLC-lite

Table 16.10 Standard function blocks

Function	Symbol	Description
Bistable function blocks		
SR	 BOOL — S1 Q1 — BOOL BOOL — R	FlipFlop, priority setting
RS	 BOOL — S Q1 — BOOL BOOL — R1	FlipFlop, priority resetting
R_TRIG	 BOOL — CLK Q — BOOL	Rising edge detection
F_TRIG	 BOOL — CLK Q — BOOL	Falling edge detection
CTU	Zähler CTU BOOL —— CU Q —— BOOL BOOL —— R INT —— PV CV —— INT	Up counter
CTD	Zähler CTD BOOL —— CD Q —— BOOL BOOL —— LD INT —— PV CV —— INT	Down counter
CTUD	Zähler CTUD BOOL —— CU QU —— BOOL BOOL —— CD QD —— BOOL BOOL —— R BOOL —— LD INT —— PV CV —— INT	Up-down counter
Timer		
TP	TP BOOL— IN Q — BOOL TIME — PT ET — TIME	IN Q ET Timer, Puls Timer, pulse
TON	TON BOOL— IN Q — BOOL TIME — PT ET — TIME	IN Q ET Timer, Einschaltverzögerung Timer, switch-on delay
TOF	TOF BOOL— IN Q — BOOL TIME — PT ET — TIME	IN Q ET Timer, Ausschaltverzögerung Timer, switch-off delay

16.2   Elements of the Instruction List Language (IL)

The statement list is composed of a sequence of statements. The instructions are structured in this way, although not every instruction must contain all of these elements:

```
Label: Operator Operand Comment
```

The following Table 16.11 lists the operators defined for the statement list. A large part of them corresponds to the standard functions mentioned in Table 16.9 and thus makes them available in the IL.

Table 16.11 Language elements Instruction list

IL			
Operator	Modifier	Operand type	Meaning
LD	N	*all*[a]	Sets the Current Result (CR) equal to the operand
ST	N	*all*[a]	Saves the CR to the operand address
S		BOOL	Sets operand to 1 if current result is TRUE
R		BOOL	Resets operand to 0 if CR is TRUE
AND	N (	ANY_BIT[a]	Bitwise AND
OR	N (	ANY_BIT[a]	Bitwise OR
XOR	N (	ANY_BIT[a]	Bitwise Exclusive OR
ADD	(	ANY_NUM, TIME	Addition
SUB	(	ANY_NUM, TIME	Subtraction
MUL	(	ANY_NUM, TIME	Multiplication
DIV	(	ANY_NUM, TIME	Division
GT	(	*all*	Comparison: >
GE	(	*all*	Comparison: >=
EQ	(	*all*	Comparison: =
NE	(	*all*	Comparison: <>
LE	(	*all*	Comparison: <
LT	(	*all*	Comparison: <=
JMP	C N	LABEL	Jump to the LABEL
CAL	C N	NAME	Call function module NAME[b]
RET	C N		Return from function or FB
)			Processing of the deferred operation

[a]When using the modifier N, the data type must be BOOL
[b]Parameters are transferred to the function block by loading/saving the input parameters as described in the front part of this book. PLC-lite does not provide for the transfer of parameters in a list of input parameters. Example: CAL_CTR_(CU:=Clock,_PV:=Input)

16.3 Keywords

Keywords (Table 16.12) are protected and must not be used for identifiers. Particular attention should be paid to this: The identifiers of the inputs and outputs of the standard function blocks are also protected keywords.

The names of the input and output parameters of the standard function blocks (formal parameters) can be found in the pictures above. These names must not be used for self-defined FBs.

Table 16.12 Keywords

Keyword
ACTION ... END_ACTION
ARRAY ... OF
AT
CASE ... OF ... ELSE ... END_CASE
CONFIGURATION ... END_CONFIGURATION
CONSTANT
EN, ENO
EXIT
FALSE
F_EDGE
FOR ... TO ... BY ... DO ... END_FOR
FUNCTION ... END_FUNCTION
FUNCTION_BLOCK ... END_FUNCTION_BLOCK
IF ... THEN ... ELSIF ... ELSE ... END_IF
IMPLEMENTS, EXTENDS
INITIAL_STEP ... END_STEP
INTERFACE ... END_INTERFACE
METHOD ... END_METHOD
PROGRAM ... WITH
PROGRAM ... END_PROGRAM
R_EDGE
READ_ONLY, READ_WRITE
REPEAT ... UNTIL ... END_REPEAT
RESOURCE ... ON ... END_RESOURCE
RETAIN, NON_RETAIN
RETURN
STEP ... END_STEP
STRUCT ... END_STRUCT

(continued)

Table 16.12   (continued)

Keyword
TASK
THIS, SUPER
TRANSITION ... FROM ... TO ... END_TRANSITION
TRUE
TYPE ... END_TYPE
VAR ... END_VAR
VAR_INPUT ... END_VAR
VAR_OUTPUT ... END_VAR
VAR_IN_OUT ... END_VAR
VAR_TEMP ... END_VAR
VAR_EXTERNAL ... END_VAR
VAR_ACCESS ... END_VAR
VAR_CONFIG ... END_VAR
VAR_GLOBAL ... END_VAR
WHILE ... DO ... END_WHILE
WITH
Data type names (see Table 16.5)
Names of the standard functions (see Table 16.9)
Names of the standard function blocks: (see Table 16.10)
CTD, CTU, CTUD, F_TRIG, RS, R_TRIG, SR, TOF, TP, TON
Identifier of the standard function block inputs/outputs:
CD, CLK, CU, CV, ET, IN, LD, PV, Q, Q1, R, R1, S, S1
Operators of the IL language (see Table 16.11)
ST language operators:
NOT, MOD, AND, XOR, OR

Terms English – German

17

Abstract

Terms English – German.

Absolute time	absolute Zeit
Access path	Zugriffspfad
Action	Aktion
Action block	Aktionsblock
Agitator	Mischer, Rührer
Aggregate	Aggregat
Argument	Argument
Array	Feld
Assignment	Zuweisung
Based number	basisbezogene Zahl
Bistable function block	bistabiler Funktionsbaustein
Bit string	Bitfolge
Boiler ·	Kessel
Body	Rumpf
Call	Aufruf
Character string	Zeichenfolge
Clock	Takt(−impuls)
Cluster	Bündel (math.)
Comment	Kommentar
Compare	vergleichen
Compile	kompilieren

(continued)

H.-J. Adam, M. Adam, *PLC Programming in Instruction list according to IEC
61131-3*, https://doi.org/10.1007/978-3-662-65254-1_17

(continued)

Configuration	Konfiguration
Cooler	Kühlung
Counter	Zähler
Counter function block	Zähler-Baustein
Data type	Datentyp
Date and time	Datum und Uhrzeit
Declaration	Deklaration
Delimiter	Begrenzungszeichen
Dice	(Spiel-)Würfel
Digit	Ziffer, Zahl
Direct representation	direkte Darstellung
Discriminate (between)	unterscheiden (zwischen)
Double word	Doppelwort
Down	abwärts
Evaluation	Auswertung
Execution control element	Element zur Ausführungssteuerung
Falling edge	fallende Flanke
Flash	Blitz, Blinker
Function	Funktion
Function block instance	Funktionsbaustein-Instanz
Function block type	Funktionsbaustein-Typ
Function block diagram	Funktionsbaustein-Sprache
Generic data type	allgemeiner Datentyp
Global scope	globaler Geltungsbereich
Global variable	globale Variable
Hierarchical addressing	hierarchische Adressierung
Identifier	Bezeichner
Initial value	Anfangswert
Input	Eingang
Input parameter	Eingangsparameter
Instance	Instanz
Instantiation	Instanziierung
Integer literal	ganzzahliges Literal
Invocation	Aufruf
Keyword	Schlüsselwort
Label	Marke
Language element	Sprachelement
Literal	Literal
Local scope	lokaler Geltungsbereich
Logical location	logischer Speicherort
Long real	lange Realzahl
Long word	Langwort
Memory	Datenspeicher
Mixer	Mischer, Rührer
Named element	bezeichnetes Element

(continued)

(continued)

Off/on-delay timer function block	Aus/Einschaltverzögerung
Operand	Operand
Operator	Operator
Output	Ausgang
Output parameter	Ausgangsparameter
Overloaded	überladen
PII (process image inputs)	PAE (Prozess-Abbild der Eingänge)
PIO (process image outputs)	PAA (Prozess-Abbild der Ausgänge)
Power flow	Stromfluß
Program	Programmieren, Programm
Program organisation unit POU	Programm-Organisations-Einheit POE
Push	Drücken
Real literal	reelles Literal
Reset	Zurücksetzen
Resource	Resource
Retentive data	Gepufferte Daten
Return	Rücksprung
Rising edge	steigende Flanke
Scope	Geltungsbereich
Semantics	Semantik
Semigraphic representation	Semigraphische Darstellung
Single data element	Einzel-Datenelement
Step	Schritt
Structured data type	Strukturierter Datentyp
Subscripting	Indizierung
Symbolic representation	Symbolische Darstellung
Switch	Schalter
Tank	Kessel, Messgefäß
Task	Task, Aufgabe
Time	Zeit
Time literal	Zeitliteral
Transition	Transition, Übergang
Unsigned integer	Vorzeichenlose ganze Zahl
Up	Aufwärts
Value	Wert
Vessel	Kessel, Messgefäß
Wired or	Verdrahtetes ODER

References

John, K.-H., Tiegelkamp, M.: SPS-Programmierung mit IEC 61131-3 – Konzepte und Programmiersprachen, Anforderungen an Programmiersysteme, Entscheidungshilfen. Springer, Heidelberg (2009)

Norm DIN EN 61131-1: Speicherprogrammierbare Steuerungen – Teil 1: Allgemeine Informationen (IEC 611311:2003); Deutsche Fassung EN 611311:2003, Beuth, Berlin (Ausgabedatum: 2004–03)

Norm SN EN 61131-1; IEC 61131-1:2003:2003: Speicherprogrammierbare Steuerungen – Teil 1: Allgemeine Informationen (IEC 61131-1:2003), Beuth, Berlin (Ausgabedatum: 2003)

Norm DIN EN 61131-3: Speicherprogrammierbare Steuerungen – Teil 3: Programmiersprachen (IEC 61131-3:2003); Deutsche Fassung EN 61131-3:2003, Beuth, Berlin (Ausgabedatum: 2003–12)

Norm DIN EN 61131-3 Beiblatt 1: Speicherprogrammierbare Steuerungen – Leitlinien für die Anwendung und Implementierung von Programmiersprachen für Speicherprogrammierbare Steuerungen, Beuth, Berlin (Ausgabedatum: 2005–04)

Normentwurf DIN IEC 61131-3: Speicherprogrammierbare Steuerungen – Teil 3: Programmiersprachen; Englische Fassung (IEC 65B/725/CD:2009), Beuth, Berlin (Ausgabedatum: 2009–12)

Norm DIN EN 61131-5: Speicherprogrammierbare Steuerungen – Teil 5: Kommunikation (IEC 61131-5:2000); Deutsche Fassung EN 61131-5:2001, Beuth, Berlin (Ausgabedatum: 2001–11)

Norm DIN EN 60848: GRAFCET – Spezifikationssprache für Funktionspläne der Ablaufsteuerung (IEC 60848:2002); Deutsche Fassung EN 60848:2002, Beuth, Berlin (Ausgabedatum: s–12)

H.-J. Adam, M. Adam, *PLC Programming in Instruction list according to IEC 61131-3*, https://doi.org/10.1007/978-3-662-65254-1

Index

© The Author(s), under exclusive license to Springer-Verlag GmbH, DE, part of
Springer Nature 2022
H.-J. Adam, M. Adam, *PLC Programming in Instruction list according to IEC 61131-3*, https://doi.org/10.1007/978-3-662-65254-1

Printed in the United States
by Baker & Taylor Publisher Services